Springer Theses

Recognizing Outstanding Ph.D. Research

Aims and Scope

The series "Springer Theses" brings together a selection of the very best Ph.D. theses from around the world and across the physical sciences. Nominated and endorsed by two recognized specialists, each published volume has been selected for its scientific excellence and the high impact of its contents for the pertinent field of research. For greater accessibility to non-specialists, the published versions include an extended introduction, as well as a foreword by the student's supervisor explaining the special relevance of the work for the field. As a whole, the series will provide a valuable resource both for newcomers to the research fields described, and for other scientists seeking detailed background information on special questions. Finally, it provides an accredited documentation of the valuable contributions made by today's younger generation of scientists.

Theses are accepted into the series by invited nomination only and must fulfill all of the following criteria

- They must be written in good English.
- The topic should fall within the confines of Chemistry, Physics, Earth Sciences, Engineering and related interdisciplinary fields such as Materials, Nanoscience, Chemical Engineering, Complex Systems and Biophysics.
- The work reported in the thesis must represent a significant scientific advance.
- If the thesis includes previously published material, permission to reproduce this must be gained from the respective copyright holder.
- They must have been examined and passed during the 12 months prior to nomination.
- Each thesis should include a foreword by the supervisor outlining the significance of its content.
- The theses should have a clearly defined structure including an introduction accessible to scientists not expert in that particular field.

More information about this series at http://www.springer.com/series/8790

Dmytro Iatsenko

Nonlinear Mode Decomposition

Theory and Applications

Doctoral Thesis accepted by
Lancaster University, UK

Springer

Author
Dr. Dmytro Iatsenko
Department of Physics
Lancaster University
Lancaster
UK

Supervisors
Prof. Aneta Stefanovska
Department of Physics
Lancaster University
Lancaster
UK

Prof. Peter V.E. McClintock
Department of Physics
Lancaster University
Lancaster
UK

ISSN 2190-5053 ISSN 2190-5061 (electronic)
Springer Theses
ISBN 978-3-319-20015-6 ISBN 978-3-319-20016-3 (eBook)
DOI 10.1007/978-3-319-20016-3

Library of Congress Control Number: 2015941872

Springer Cham Heidelberg New York Dordrecht London

Printed on acid-free paper

Springer International Publishing AG Switzerland is part of Springer Science+Business Media
(www.springer.com)

Parts of this thesis have been published in the following journal articles:

1. D. Iatsenko, P. V. E. McClintock and A. Stefanovska, Nonlinear Mode Decomposition: a new noise-robust, adaptive, decomposition method, submitted for publication [preprint—arXiv:1207.5567]
2. D. Iatsenko, P. V. E. McClintock and A. Stefanovska, Glassy states and superrelaxation in populations of coupled phase oscillators, *Nature Comm.* **5**, 4118 (2014) [Open Access]
3. D. Iatsenko, S. Petkoski, P. V. E. McClintock and A. Stefanovska, Stationary and traveling wave states of the Kuramoto model with an arbitrary distribution of frequencies and coupling strengths, *Phys. Rev. Lett.* **110**, 064101 (2013) [preprint—arXiv:1208.4561]
4. D. Iatsenko, A. Bernjak, T. Stankovski, Y. Shiogai, P. J. Owen-Lynch, P. B. M. Clarkson, P. V. E. McClintock and A. Stefanovska, Evolution of cardiorespiratory interactions with age, *Phil. Trans. R. Soc. A* **371**, 20110622 (2013)
5. D. Iatsenko, P. V. E. McClintock and A. Stefanovska, Linear and synchrosqueezed time-frequency representations revisited: Overview, standards of use, resolution, reconstruction, concentration, and algorithms, *Dig. Signal Proc. (in press)*, doi:10.1016/j.dsp.2015.03.004 (2015)
6. D. Iatsenko, P. V. E. McClintock and A. Stefanovska, On the extraction of instantaneous frequencies from ridges in time-frequency representations of signals, submitted for publication [preprint—arXiv:1310.7276]
7. D. Iatsenko, A. Stefanovska and P. V. E. McClintock, Comment on "Inference with minimal Gibbs free energy in information field theory", *Phys. Rev. E* **85**, 033101 (2012)
8. S. Petkoski, D. Iatsenko, L. Basnarkov and A. Stefanovska, Mean-field and mean-ensemble frequencies of a system of coupled oscillators, *Phys. Rev. E* **87**, 032908 (2013) [preprint—arXiv:1302.7164]
9. D. Iatsenko, P. V. E. McClintock and A. Stefanovska, Impaired coordination of cardiovascular oscillations in ageing and treated hypertension, in preparation

Miscellaneous

- Nonlinear Mode Decomposition Toolbox for MatLab and many other programs, freely available at http://www.physics.lancs.ac.uk/research/nbmphysics/diats/nmd/ and http://www.physics.lancs.ac.uk/research/nbmphysics/diats/tfr/. All codes are supplemented with a detailed documentation—and many of them with video instructions and examples.

Presentations

- SET for Britain 2014 (poster presentation in the UK House of Commons, 17 March 2014)
- Dynamics Days US 2013 (poster presentation, 5 January 2013)
- Lancaster Christmas Conference 2012 (poster presentation, 17 December 2012)

Awards

- Dean's Award for Ph.D. excellence (first year category), Faculty of Science and Technology, Lancaster University (2012)
- Best International Ph.D. student in Physics Department, Lancaster University (2012)
- Best Ph.D. student in Physics Department, Lancaster University (2014)

Supervisor's Foreword

In a quite general sense science can be perceived as dealing with two universal questions:

- How things work, on all scales, from the macroscopic down to the most microscopic level possible? and
- Why things sometimes do not work, or do not work correctly?

But can we really understand why things sometimes do not work by knowing about their function when their structure is normal? For example in medicine, can we diagnose diseases by analysing and mapping genes, the present day atomic level for living systems? Can we sometimes map a given disease onto particular sets of genes, and then replace those genes to cure the disease? It would have been wonderful if this scenario was possible—but living systems are altogether too complex for such a picture to apply except perhaps in a few exceptional cases.

With complex systems, where the whole is more than the sum of the parts [1], in addition to understanding the role of each component, one needs to understand how they interact. *Interaction* is a property of the functioning of a system. While the structure is relatively easy to establish—with today's advanced technology, effective imaging can be done at and below the nano-level—the function is usually much more difficult to decipher. This is because of two extremes inherent in every function:

(a) When only one entity is being considered, and it has nothing else to interact with, there is no function. Function is meaningful only when interactions exist. Of course, the entity in question is probably made up of more micro-scopic elements, so that it may have internal function, and the boundaries are being pushed continuously with the advent of new technology. So, one speaks of the currently accessible microscopic level.
(b) But as soon as we know interactions to exist, we face other questions: how should the basic functional units be defined? Furthermore, what is the contribution of the interactions and how are they mediated?

Hence, despite the breath-taking speed of progress in science and technology, or perhaps as a consequence, we face the problem of how best to make use of these advances. Ideally, one can expect that the macroscopic-down-to-microscopic scale approach (also known as the top-down approach) will include the study of inter-actions and provide details of how things work at every level of complexity. But in the top-down approach, the interactions are often overlooked.

In the bottom-up approach everything is observed and taken account of simultaneously. On the other hand one can expect that the complex systems approach—when all parts are left to exist without greatly perturbing them—could bring huge advances, especially in understanding why either natural or man-made systems go wrong. For this approach to be successful one expects that the function of the system in question will be recorded in some appropriate way, and preferably in a way that is non-perturbative (non-invasive) over the relevant period, and recording simultaneously at as many as possible levels of complexity. However, this brings yet another set of problems: we can easily end up with too much information which, in turn, brings huge difficulties of interpretation. Two factors are crucial here:

(i) The availability of basic knowledge about all the parts, and about how they interact.
(ii) The availability of good methods for making use of the recorded data. These include good physical models and theories for non-isolated and complex systems, as well as good methods for the inverse approach, known as time-series analysis, or signal analysis or data analysis–depending on whether physicists or mathematicians or, more recently, computer scientists, are dominating the enterprise.

The step formulated under (ii) is a fundamental problem of decomposition, and it is relevant to practically every area of science: from biology to astrophysics, from micro-economics to sociology, from cell dynamics to brain dynamics, from nature-made to man-made systems.

Although a number of methods have been proposed for dealing with many the obstacles that arise when tackling the problem of decomposition, the major unsolved problem to date has been to develop a form of decomposition that takes explicit account of nonlinearitiy. Nonlinearities can be present both in the basic functioning units as well as in their interactions. Dima has confronted this problem directly, and has proposed a new method of decomposition which he has named *Nonlinear Mode Decomposition* (NMD).

Of course, every method of decomposition assumes the existence of determinism. Complex systems with a large deterministic contribution are often oscillatory and, moreover, they often have time-varying characteristic oscillation frequencies. Therefore, the central part of the work is detection of the "ridges" of the basic modes and of high harmonics of nonlinear, time varying, oscillatory processes. Dima has made excellent use of a recently-proposed method for detecting the high harmonics of noisy time-variable oscillatory processes, based on mutual information and surrogate testing [2].

His work is very systematic and detailed. It not only presents the NMD method, but also reviews critically the methods available for time-frequency analysis, including the short-time Fourier transform and wavelet transform. It introduces key concepts, such as the analytic signal representation, instantaneous frequency, instantaneous amplitude, and how to detect them from real data. It also introduces ridge curves and discusses methods for their extraction. Finally, it introduces NMD itself, which consists of three main steps:

(a) Obtain a time-frequency representation of the signal and extract the time-evolution of the fundamental harmonics (basic frequency) of the nonlinear mode.
(b) Select candidates for all high harmonics and identify the true harmonics.
(c) Reconstruct the full nonlinear mode by summing together all the true harmonics; subtract it from the signal; and iterate the procedure on the residual until a pre-set stopping criterion is met.

Dima has illustrated the application of the method, not only with several numerical examples, but also on real signals related to cardiovascular and brain dynamics. Given that nearly all real systems of an oscillatory character have imperfect clocks, so that their characteristic frequencies are time-variable, the method can be expected to be useful in practically all areas of science. The work also provides an excellent reference for those who would like to gain an in-depth understanding of recent advances in time-frequency analysis, quite generally.

Lancaster
June 2015

Prof. Aneta Stefanovska

References

1. W. Heisenberg, *Physics and Beyond: Encounters and Conversations*, (Harper and Row, New York, 1971)
2. L W Sheppard, A Stefanovska, and P V E McClintock, "Detecting the harmonics of oscillations with time-variable frequencies", *Phys. Rev. E* **83**, 016206 (2011)

Abstract

This thesis introduces a new adaptive decomposition method—Nonlinear Mode Decomposition (NMD)—which decomposes a given signal into a set of physically meaningful oscillations within any waveform, simultaneously removing the noise. It is based on the powerful combination of two elements. First, time-frequency analysis techniques with the adaptive parameter choice make the method extremely noise-robust. Secondly, surrogate data tests are used to identify interdependent oscillations and to distinguish deterministic from random activity.

The theory of linear time-frequency representations, which represent the foundation of NMD, is first reviewed and advanced, with emphasis being placed on its practically relevant aspects. Techniques for extracting harmonic oscillations with time-varying amplitudes and frequencies from the signal's time-frequency representation are then developed. By combining these techniques with additional procedures devised for distinguishing the retrieved oscillations from noise and for the recovery of their full waveforms, the NMD is finally formed.

The performance of the method is illustrated on both simulated and real signals, and its qualitative and quantitative superiority over the other existing approaches (such as (ensemble) empirical mode decomposition, Karhunen-Loève expansion and independent component analysis) is shown. In particular, NMD is applied for the decomposition of human blood flow signals and, based on properties of the recovered oscillations for different subject groups, certain aspects of cardiovascular ageing and (treated) hypertension are revealed. Furthermore, applications of the method for removing the measurement artefacts from a single electroencephalogram recording, and for distinguishing different kinds of systems, are also demonstrated. These examples, however, represent only a few out of many possible uses of NMD, which can be applied routinely to a diversity of signals coming from various scientific areas (geophysics, finance, life sciences, etc.). The necessary MATLAB codes for running NMD are freely available for download.

Acknowledgments

Above all, I would like to thank my principal supervisor, Prof. Aneta Stefanovska, for the continuous support of my Ph.D. study and research, and for the great attitude, care, patience and help in all aspects. It is because of her I have been given such a fantastic opportunity to discover this exciting area, to learn, to develop and to grow.

I am deeply grateful to my second supervisor, Prof. Peter McClintock, for his invaluable help and advice, for his enormous efforts and patience in correcting my writing and for the many joyful discussions we had.

My sincere gratitude also goes to my colleagues with whom I have worked closely on some papers, namely Dr. Spase Petkoski, Dr. Alan Bernjak and Dr. Tomislav Stankovski. Special thanks go to Dr. Phil Clemson, Gemma Lancaster and Dr. Yevhen Suprunenko for valuable discussions and useful comments on the thesis. Additionally, although I have not been able to work more closely with them yet, I am grateful to Valentina Ticcinelli and Will Gibby for creating such a nice atmosphere in the group.

Besides academic acknowledgements, I would like to thank my family and friends, and especially my parents, for their encouragement and continual support. Extra thanks go to Hyleys and Ahmad brands for making strong green teas that I have used frequently to boost my mental performance.

Contents

1 Introduction 1
1.1 Preamble 1
1.2 Outline 4
References 5

2 Linear Time-Frequency Analysis 7
2.1 AM/FM Components and the Analytic Signal 8
2.2 Time-Frequency Representations (TFRs) 11
2.2.1 Windowed Fourier Transform (WFT) 12
2.2.2 Wavelet Transform (WT) 14
2.2.3 Difference Between the WFT and the WT 17
2.3 Time-, Frequency- and Time-Frequency Resolution 19
2.3.1 General Formulation 19
2.3.2 Classical Definitions and Their Flaws 22
2.3.3 Notion of the Window/Wavelet ϵ-Support 23
2.3.4 Reconsidered Definitions 25
2.4 Practical Issues 27
2.4.1 Signal Preprocessing 27
2.4.2 Frequency Discretization 30
2.4.3 Boundary Effects and Padding 32
2.4.4 TFR Frequency Range 38
2.5 Summary 39
References 40

3 Extraction of Components from the TFR 43
3.1 Identification of the Components 43
3.1.1 Ridge Curve and Its Extraction 43
3.1.2 Main Procedure 46
3.2 Estimation of the Component's Parameters 48
3.2.1 Ridge Reconstruction 48
3.2.2 Direct Reconstruction 49

3.2.3 Difference Between the Two Estimates 52
3.2.4 Adaptive Choice of the Reconstruction Method 53
3.3 Summary 56
References 57

4 Nonlinear Mode Decomposition (NMD) 59
4.1 Basic Idea and Ingredients 60
4.1.1 Extraction of the Fundamental Harmonic 61
4.1.2 Harmonics: Extracting the Candidates 61
4.1.3 Harmonics: Identifying the True Ones 61
4.1.4 Harmonics: Practical Issues 64
4.1.5 Stopping Criterion 66
4.2 Improvements 69
4.2.1 Adaptive Representation of the Harmonics 69
4.2.2 Improved Reconstruction of Nonlinear Modes 74
4.3 Parameters and Their Choice 75
4.3.1 Resolution Parameter f_0 75
4.3.2 TFR Type: WFT or WT? 77
4.3.3 Reconstruction Method: Direct or Ridge? 79
4.3.4 Other Parameters 79
4.4 Summary 80
References 81

5 Examples, Applications and Related Issues 83
5.1 Simulated Examples 83
5.1.1 Example 1 83
5.1.2 Example 2 86
5.2 Real Applications 89
5.2.1 Decomposing Human Blood Flow Signals 89
5.2.2 Clinical Application: Cardiac Waveform Study 92
5.2.3 Removing Cardiac Artifacts from a Human EEG Signal 95
5.2.4 Other Applications 97
5.3 Physical Interpretation and System Classification 98
5.3.1 "Decomposability" of Different Systems 99
5.3.2 Chronotaxic Systems and Their Identification 105
5.4 Limitations and Possible Issues 109
References 110

6 Conclusion 113
6.1 Summary 113
6.2 Original Contributions 114
6.3 Future Perspectives 115
References 115

7 Appendix: Useful Information and Derivations 117
7.1 Errors of the Analytic Estimates 117
7.2 Window and Wavelet Functions and Their Properties 119
7.2.1 Window Functions 119
7.2.2 Wavelet Functions 122
7.3 Derivation of the Direct Reconstruction Formulas 124
7.4 Forecasting Model for Predictive Padding 126
7.5 Step-by-step WFT and WT Algorithms 129
7.5.1 WFT $G_s(\omega_k, t_n)$ 130
7.5.2 WT $W_s(\omega_k, t_n)$ 131
7.6 Fast $O(N)$ Algorithm for Path Optimization 132
References 135

Abbreviations and Symbols

All abbreviations and notation will be gradually introduced in the main text, but for convenience they are all listed below together with the references to the places in the text where one can find their precise meaning and related discussion.

Abbreviations

BCG	Ballistocardiogram (artifacts)
ECG	Electrocardiogram
EEG	Electroencephalogram
EEMD	Ensemble Empirical Mode Decomposition
EMD	Empirical Mode Decomposition
FT	Fourier Transform
FFT	Fast Fourier Transform (algorithm)
ICA	Independent Component Analysis
IFFT	Inverse Fast Fourier Transform (algorithm)
LDF	Laser-Doppler flowmetry (the method for measuring skin blood flow, see Sect. 5.2.1)
NM	Nonlinear Mode (see (4.1) and its discussion)
NMD	Nonlinear Mode Decomposition (see Chap. 4)
PCA	Principal Component Analysis
TFR	Time-Frequency Representation (includes WFT, WT and many others, but in this work only the former two are considered)
TFS	Time-Frequency Support (see Sect. 3.2.2)
WFT	Windowed Fourier Transform (see Sect. 2.2.1)
WT	Wavelet Transform (see Sect. 2.2.2)

Terminology

AM/FM component, or simply component	Sinusoidal oscillation with amplitude and/or frequency modulation, i.e. the function of time t of the form $x(t) = A(t)\cos\phi(t)$ with $A(t) > 0, \phi'(t) > 0, \forall t$ (see Sect. 2.1 for a more detailed discussion)
Tone	AM/FM component of constant amplitude and frequency, i.e. a simple sine $x(t) = A\cos(\nu t + \varphi)$
Gaussian window	Window function of the form (2.12)
Morlet wavelet	Wavelet function of the form (2.18)
Lognormal wavelet	Wavelet function of the form (2.19)
Time resolution of the TFR	The reciprocal of the minimum time lag between the two time events for which they can both be represented reliably in the TFR (see Sect. 2.3)
Frequency resolution of the TFR	The reciprocal of the minimum frequency difference between the two frequency events (tones) for which they can both be represented reliably in the TFR (see Sect. 2.3)
Joint time-frequency resolution of the TFR	The reciprocal of the minimum time-frequency area where the interference between two time events and between two frequency events is simultaneously small in the TFR (see Sect. 2.3)
Nonlinear Mode (NM)	Amplitude and/or frequency modulated oscillation which, in contrast to simple AM/FM component, is allowed to have a complex waveform. Mathematically, it is defined as a function of time t of the form $c(t) = A(t)\sum_h a_h \cos(\phi(t) + \varphi_h)$ with $A(t) > 0, \phi'(t) > 0, \forall t$ (see (4.1) and its discussion for more details)
Harmonic	AM/FM component which represents a part of some nonlinear mode (4.1); harmonic with the lowest frequency among those corresponding to the same NM is referred to as its fundamental (or first) harmonic (see (4.1) and its discussion for more details)

Main Notation

$\hat{f}(\xi)$	Fourier Transform of $f(t)$, see (2.2)
$f^{\pm}(t)$	Positive/negative frequency part of $f(t)$, see (2.2)
$\langle f(t)\rangle$	Time average of $f(t)$, see (2.2)
$std[f(t)]$	Standard deviation of $f(t)$, see (2.2)
$f^{*}(t)$	Complex conjugate of $f(t)$
$c.c.$	Complex conjugate of the preceding expression
$\mathrm{Re}[x], \mathrm{Im}[x]$	Real and imaginary parts of x
$s^{a}(t)$	Analytic signal (2.3), which is twice the signal's positive frequency part $s^{a}(t) = 2s^{+}(t)$. If the original signal is real, as is assumed in this work, then $s(t) - \langle s(t)\rangle = \mathrm{Re}[s^{a}(t)]$
$sign(x)$	Sign function: $sign(x > 0) = 1$, $sign(x < 0) = -1$, $sign(0) = 0$
$J_n(x)$	nth order Bessel function of the first kind
$I_n(x)$	nth order modified Bessel function of the first kind
$\Gamma(a)$	Gamma function $\Gamma(n) \equiv \int_0^{\infty} x^{a-1}e^{-x}dx$ $(= (a-1)!$ if $a \in \mathbb{N})$
$erf(x)$	Gauss error function $\mathrm{erf}(x) \equiv \frac{2}{\sqrt{\pi}}\int_0^{x} e^{-u^2}du$
$n_G(\varepsilon)$	Number of standard deviations of the Gaussian distribution within which its $(1-\varepsilon)$ part is contained, i.e. $\frac{\int_{-n_G}^{n_G} e^{-u^2/2}du}{\int_{-\infty}^{\infty} e^{-u^2/2}du} = \mathrm{erf}(n_G/\sqrt{2}) = 1-\varepsilon$. For example, $n_G(0.05) \approx 2$, $n_G(0.01) \approx 2.5$, $n_G(0.001) \approx 3.3$
$G_s(t,\omega), W_s(t,\omega)$	WFT (2.8) and WT (2.13) of the considered signal $s(t)$, respectively
$g(t), \hat{g}(\xi)$	Time domain and frequency domain forms of the window function used for computation of the WFT (2.8), respectively
$\psi(t), \hat{\psi}(\xi)$	Time domain and frequency domain forms of the wavelet function used for computation of the WT (2.13), respectively
ω_g, ω_ψ	Window and wavelet peak frequencies: $\omega_g \equiv \operatorname{argmax}_{\omega} \lvert\hat{g}(\omega)\rvert$, $\omega_\psi \equiv \operatorname{argmax}_{\omega} \lvert\hat{\psi}(\omega)\rvert$. It is assumed that $\omega_g = 0$
C_g, C_ψ	Integration constants defined in (2.11) and (2.17), respectively
$\overline{\omega}_g, D_\psi$	Integration constants defined in (7.9) and (7.10), respectively
f_0	Window/wavelet resolution parameter which determines the trade-off between its time and frequency resolutions (in the sense that increasing f_0 increases frequency resolution but decreases time resolution). The way it is introduced for different window and wavelet functions is described in Appendix 7.2
$R_g(\omega), P_g(\tau)$	Quantitative measures of the area below $\hat{g}(\xi < \omega)$ and $g(t < \tau)$, respectively, as defined in (2.26)

$R_\psi(\omega), P_\psi(\tau)$	Quantitative measures of the area below $\hat{\psi}(0<\xi<\omega)$ and $g(t<\tau)$, respectively, as defined in (2.27)
$\xi_{1,2}(\varepsilon)$, $\tau_{1,2}(\varepsilon)$	ε-supports of the window/wavelet function in time and frequency, defined in (2.26) for windows and in (2.27) for wavelets; whether it refers to the former or to the latter is always clear from the context
$H_s(\omega, t)$	Used to denote both $G_s(\omega, t)$ and $W_s(\omega, t)$ in expressions which are the same for both WFT and WT
$[\omega_{\min}, \omega_{\max}]$	Frequency range for which the currently considered TFR is calculated
$\omega_p(t)$	Ridge curve of some component in the currently considered TFR (see Sect. 3.1.1)
$[\omega_-(t), \omega_+(t)]$	Time-frequency support of some component in the currently considered TFR (see Sect. 3.2.2)
a_h, φ_h	Amplitude ratios and phase shifts of harmonics for the currently considered nonlinear mode (4.1)
$\eta_W(t)$	Gaussian white noise of unit variance
$\eta_B(t)$	Brownian noise of unit variance, constructed as a (normalized) cumulative sum of the Gaussian white noise signal: $\eta_B(t_n) \sim \sum_{i=0}^{n-1} \eta_W(i\Delta t)$

Assumptions and Conventions

- Where undefined, the integrals are taken over $(-\infty, \infty)$ or, in practice, over the full range of the corresponding variable. For example, given a signal $s(t)$, one has $\int s(t)dt \equiv \int_{-\infty}^{\infty} s(t)dt$ theoretically and $\int s(t)dt \equiv \int_0^T s(t)dt$ practically, with T denoting the overall time duration of the signal.
- All TFRs are assumed to be computed for a real signal $s(t)$. Additionally, without loss of generality it is always assumed that $\underset{\xi}{\operatorname{argmax}} |\hat{g}(\xi)| = 0$, i.e. that the window function used for calculating the WFT (2.8) is peaked at zero in the frequency domain (see Sect. 2.2.1).
- In all examples, unless specified otherwise, the WFT (WT) is calculated using Gaussian window (2.12) (lognormal wavelet (2.19)) with $f_0 = 1$. The frequency axis is discretized as $\omega_k = (k - k_0)\Delta\omega$ for the WFT, and as $\omega_k/2\pi = 2^{(k-k_0)/n_v}$ for the WT, with the discretization parameters $\Delta\omega$ and n_v being selected according to the criteria (2.34) with $N_b = 10$ (yielding $\Delta\omega/2\pi = 0.02$ and $n_v = 33$ for the Gaussian window and lognormal wavelet with $f_0 = 1$, respectively). To reduce boundary effects in the TFRs (see Sect. 2.4.3), the predictive padding is used by default, with the number of padded values being calculated according

to (2.36); in the majority of cases, however, only the “distortion-free” TFR parts (i.e. those lying within the corresponding cones-of-influence (2.39) with $\varepsilon = 0.01$) are presented.

- To simplify all expressions, the circular frequencies—in rad/s—are mainly used, denoted by ω, ν, ξ (with additional subscripts or superscripts); to convert these frequencies to Hz, one needs to divide them by 2π. Note, however, that the signal’s sampling frequency f_s is always taken to be in Hz.

Chapter 1
Introduction

1.1 Preamble

Oscillatory systems are abundant in nature, surrounding us on all sides. Consider, for example, the propagating vibrations of a medium that are perceived as sound, the rhythmic heart pulses pushing blood through the arteries and veins, the repeated movements of air through the lungs known as breathing, the diurnal (day-night) cycle, and periodic climate changes. All of these are examples of oscillations. It is therefore beyond doubt that to understand better the world we live in, or at least to create a more efficient view of it—which is the main aim of any science—one needs to study such oscillatory processes.

In practice, real systems are investigated by analysing the observable data that they generate, which is most commonly presented in the form of time series (i.e. discrete time signals). These consist of sequences of data points, each having its own reference time. Simple examples of such time series signals would be an hourly measured solar flux intensities, or the body temperature measured once per minute.

Due to the very abundance of oscillatory systems, measured signals are usually composed of a mixture of different oscillations with complex waveforms and time-varying amplitudes and frequencies. These oscillations, which will also be referred to as modes, contain a great deal of valuable information about the originating system. They can be used to predict earthquakes from geophysical signals [1], or to infer possible future market movements from a financial time series [2]. The oscillatory properties of the signals generated by pieces of machinery can be used to assess its functionality and to diagnose faults [3–5]; the same is true for signals generated by the human body (e.g. the electrical activity of the heart), which often have both diagnostic and prognostic value for a variety of diseases [6, 7]. Quite generally, the function of any living organism depends on a range of oscillatory processes taking place on all scales, from microscopic to macroscopic [8, 9].

The question is therefore how to gain access to the properties of the individual oscillations contained in a measured signal. Though one can often estimate certain characteristics indirectly, to gain access to all the properties of a particular mode

D. Iatsenko, *Nonlinear Mode Decomposition*,
Springer Theses, DOI 10.1007/978-3-319-20016-3_1

it should first be extracted from the signal. Usually, this task is additionally complicated by the presence of noise—an unwanted pseudorandom activity arising as the cumulative outcome of a large number of small events or weak processes (e.g. generated by unsynchronized microscopic oscillators [10]). For a reliable analysis, therefore, one needs to decompose the signal, i.e. recover the individual oscillations that are present in it, separating them both from each other and from the inevitable background of noise.

How best to accomplish this decomposition is a long-standing problem, and many different approaches have been proposed. Historically, the first approach of this kind was probably the well-known Fourier transform (FT) introduced by Joseph Fourier in 1822. It represents the signal as a linear superposition of trigonometric functions, thus decomposing it into a set of tones—oscillations with constant amplitudes and frequencies. This approach was so successful, that nearly all areas of research benefitted from it, and now Fourier analysis is widely used in acoustics, optics, quantum mechanics, econometrics, mathematics and many other disciplines.

However, it is rare for real oscillations to be strictly periodic: they are usually characterized by time-dependences of their frequency and other parameters. Thus, the heart does not beat at a constant rate, but continuously adjusts itself, depending on internal and external influences, such as the respiratory cycle [11, 12]; in turn, the respiratory cycle is itself also characterized by a time-varying period, as we breath quite irregularly, sometimes making a long deep breath and sometimes breathing fast and shallowly [12, 13]; even diurnal cycles change their period slowly due to the deceleration of the Earth's rotation [14]. The Fourier transform cannot represent such time-dependent oscillations as single entities and, instead, it decomposes them into multiple tones. The latter are therefore of purely mathematical utility and do not make physical sense, so that the Fourier transform is not usefully interpretable in such cases.

A natural way to extend Fourier analysis to the case of amplitude/frequency modulated oscillations is to analyse the FT computed for a sequence of relatively short sections of signal centered on different times: this can conveniently be effected with the aid of a sliding time window. This idea gave birth to the so-called windowed (or short-time) Fourier transform (WFT) [15–17]. Later on, it was observed that the use of adaptive windows scaled for each frequency can be advantageous in many cases, which led to the introduction of the wavelet transform (WT) [17–19]. In addition to the WFT and WT, there are numerous other ways of projecting the signal onto the time-frequency plane, with such projections generally being called *time-frequency representations* (TFRs); the Wigner-Ville [20, 21], Rihaczek [22] and Choi-Williams [23] distributions are the other well-known examples of the TFRs (see [16, 17, 19, 24–28] for a number of excellent books and reviews on the topic).

In the time-frequency plane, the oscillatory components present in the signal appear in the form of "curves", formed by the sequences of TFR amplitude peaks. Having identified these curves, one can then recover the corresponding oscillations, thereby decomposing the signal. This idea gave birth to the proposal of a variety of techniques in recent years [29–34], with great advantage of such approaches being their high noise-robustness [35, 36].

In line with the development of TFRs, there has also been an intensive activity on developing decomposition methods for multivariate time-series. The most widely-used among them are principal component analysis (PCA) [37, 38] and independent component analysis (ICA) [39, 40], as well as their numerous modifications and extensions [41, 42]. These methods assume that all signals are linear mixtures of the same scaled parts, and try to find these parts as the ones that are minimally correlated (PCA) or maximally independent (ICA). In the present work, however, consideration is restricted to univariate signals, i.e. those represented by a single time-series. To apply PCA and ICA in this case, one should first construct a multivariate signal from the original (univariate) one, which can be accomplished by using its time-shifted blocks [43]. PCA and ICA can then be applied to this collection of the original signal's parts, in which case they are called single-channel PCA (also known as Karhunen-Loève expansion) and single-channel ICA [44], respectively.

Finally, the other approach for decomposing a univariate signal into a set of oscillations with time-varying amplitudes and frequencies, empirical mode decomposition (EMD), was proposed in 1998 [45]. This method works directly in the time domain and is based on identification and interpolation of peaks and troughs in the signal. Although purely empirical and very sensitive to noise, EMD nonetheless became very popular and has already been cited more than 4500 times (according to Web of Science, while Google Scholar returns >9000 citations), a fact that reflects very well the importance of the problem being addressed. As a next step, Wu and Huang [46] proposed a modification called ensemble empirical mode decomposition (EEMD), which aims to overcome the high noise-sensitivity and other drawbacks of EMD. The basic idea behind it is to add independent realizations of white Gaussian noise to the signal and to apply EMD each time, with the final modes being obtained as the ensemble averages of the modes for each noise realization.

Each of the above-mentioned approaches, however, suffers from at least two of the following flaws:

1. The method contains user-defined parameters and is quite sensitive to their choice. These parameters cannot be adaptively chosen and there are no more-or-less universal settings.
2. The method is not noise-robust.
3. If an individual mode has a complex (nonsinusoidal) waveform, the method will decompose it into a few distinct oscillations with simpler waveforms.
4. The modes returned by the method are not always physically meaningful. For example, the method will decompose even a random signal, such as Brownian noise, into a set of oscillations.

Thus, the EMD method suffers from disadvantages 2–4, with 3 manifesting itself only in cases where the corresponding oscillation has a waveform with multiple peaks. The EEMD modification to some extent removes the disadvantage 2 (though still not making the method really noise-robust, at least in comparison with the alternative approaches), but other flaws remain. Single-channel PCA (or Karhunen-Loève expansion) and single-channel ICA suffer from drawback 1, as they both are very susceptible to the choice of time-shifts and of the number of blocks used to create the multivariate signal from the original one (as was demonstrated e.g. in [43] for Karhunen-Loève expansion); in addition, they also suffer from flaw number 4. Finally, the TFR-based methods, in addition to disadvantage 4, which is common to all current approaches, also suffer from drawback 3, because any oscillation with a non-sinusoidal waveform is typically represented by a number of curves in the time-frequency plane.

These drawbacks greatly restrict the applicability of the approaches currently in use, so that the decomposition can be carried out successfully only for a small class of signals. The present work therefore introduces a new method—*Nonlinear Mode Decomposition* (NMD)—which is intended to be free from all the disadvantages considered above. The emphasis of this work is therefore the development of NMD and its applications, which include (but are not limited to) the decomposition of blood flow signals, clinical studies of the effects of ageing and hypertension, artifact removal and system classification.

1.2 Outline

Nonlinear mode decomposition consists of many elements, and for a thorough understanding of all of them this work builds up the method block-by-block starting from its foundations. Thus, the WFT and WT, which represent the very basis of NMD, are first considered in Chap. 2. It aims to provide a detailed review of their properties and related issues, focusing on the most practically relevant aspects and not going too deeply into the mathematical theory.

Next, Chap. 3 considers the identification and reconstruction of the oscillatory components present in the signal based on its WFT or WT. It develops the techniques for tracking the oscillatory components in the time-frequency plane, and for estimating their properties. These techniques allow one to extract the constituent oscillations from the signal.

Based on theory reviewed/developed in Chaps. 2 and 3, NMD is constructed in Chap. 4, where its subprocedures, their realizations and possible improvements are developed one-by-one. The particular emphasis is put on the extraction of oscillations with complex waveforms and related issues, as well as on the determining the number of meaningful oscillations in the signal. The corresponding techniques, being combined with those developed in Chap. 3, form the final decomposition method.

The operation and performance of the method on simulated examples are illustrated in Chap. 5, and NMD is then applied to various real-life problems. Chapter 6 summarizes the work and outlines future perspectives together with the possible improvements, while Appendix provides mostly technical details that are omitted from the main text to avoid interrupting the smooth flow of ideas.

References

1. G.C. Beroza, T.H. Jordan, Searching for slow and silent earthquakes using free oscillations. J. Geophys. Res. **95**(B3), 2485–2510 (1990)
2. X. Zhang, K.K. Lai, S.-Y. Wang, A new approach for crude oil price analysis based on empirical mode decomposition. Energ. Econ. **30**(3), 905–918 (2008)
3. U. Benko, J. Petrovčič, Ð. Juričić, J. Tavčar, J. Rejec, A. Stefanovska, Fault diagnosis of a vacuum cleaner motor by means of sound analysis. J. Sound Vib. **276**(3), 781–806 (2004)
4. S. Nandi, H.A. Toliyat, X. Li, Condition monitoring and fault diagnosis of electrical motors-a review. IEEE Trans. Energy Convers. **20**(4), 719–729 (2005)
5. Z.K. Peng, F.L. Chu, Application of the wavelet transform in machine condition monitoring and fault diagnostics: a review with bibliography. Mech. Syst. Sig. Process. **18**(2), 199–221 (2004)
6. G. Parati, J.P. Saul, M. Di Rienzo, G. Mancia, Spectral analysis of blood pressure and heart rate variability in evaluating cardiovascular regulation: a critical appraisal. Hypertens. **25**(6), 1276–1286 (1995)
7. Y. Shiogai, A. Stefanovska, P.V.E. McClintock, Nonlinear dynamics of cardiovascular ageing. Phys. Rep. **488**, 51–110 (2010)
8. M.J. Costa, B. Finkenstädt, V. Roche, F. Lévi, P.D. Gould, J. Foreman, K. Halliday, A. Hall, D.A. Rand, Inference on periodicity of circadian time series. Biostatistics **14**(4), 792–806 (2013)
9. A. Goldbeter, *Biochemical Oscillations and Cellular Rhythms: the Molecular Bases of Periodic and Chaotic Behaviour* (Cambridge University Press, Cambridge, 1996)
10. L.W. Sheppard, A.C. Hale, S. Petkoski, P.V.E. McClintock, A. Stefanovska, Characterizing an ensemble of interacting oscillators: the mean-field variability index. Phys. Rev. E **87**(1), 012905 (2013)
11. J.A. Hirsch, B. Bishop, Respiratory sinus arrhythmia in humans: how breathing pattern modulates heart rate. Am. J. Physiol. Heart. C. **241**(4), H620–H629 (1981)
12. D. Iatsenko, A. Bernjak, T. Stankovski, Y. Shiogai, P.J. Owen-Lynch, P.B.M. Clarkson, P.V.E. McClintock, A. Stefanovska, Evolution of cardio-respiratory interactions with age. Phil. Trans. R. Soc. Lond. A **371**(1997), 20110622 (2013)
13. J.G. Aardweg, J.M. Karemaker, Respiratory variability and associated cardiovascular changes in adults at rest. Clin. Physiol. **11**(2), 95–118 (1991)
14. F.R. Stephenson, L.V. Morrison, G.J. Whitrow, Long-term changes in the rotation of the earth: 700 BC to AD 1980 [and discussion]. Phil. Trans. R. Soc. Lond. A **313**(1524), 47–70 (1984)
15. J.B. Allen, L. Rabiner, A unified approach to short-time fourier analysis and synthesis. Proc. IEEE **65**(11), 1558–1564 (1977)
16. L. Cohen, in *Time-Frequency Analysis*, vol. 778 (Prentice Hall PTR, New Jersey, 1995)
17. F. Hlawatsch, G.F. Boudreaux-Bartels, Linear and quadratic time-frequency signal representations. IEEE Sig. Proc. Mag. **9**(2), 21–67 (1992)
18. I. Daubechies, The wavelet transform, time-frequency localization and signal analysis. IEEE Trans. Inf. Theory **36**(5), 961–1005 (1990)
19. I. Daubechies, in *Ten Lectures on Wavelets*, vol. 61 of CBMS-NSF Regional Conference Series in Applied Mathematics, SIAM, 1992
20. J.D. Ville et al., Théorie et Applications de la Notion de Signal Analytique. Câables et transmission, **2**(1), 61–74 (1948)
21. E. Wigner, On the quantum correction for thermodynamic equilibrium. Phys. Rev. **40**(5), 749 (1932)
22. A.W. Rihaczek, Signal energy distribution in time and frequency. IEEE Trans. Inf. Theory **14**(3), 369–374 (1968)
23. H.-I. Choi, W.J. Williams, Improved time-frequency representation of multicomponent signals using exponential kernels. IEEE Trans. Acoust. Speech Sig. Process. **37**(6), 862–871 (1989)
24. P.S. Addison, *The Illustrated Wavelet Transform Handbook: Introductory Theory and Applications in Science, Engineering Medicine and Finance* (IOP Publishing, Bristol, 2002)

25. B. Boashash, *Time Frequency Signal Analysis and Processing*. (Elsevier, Amsterdam, 2003)
26. F. Hlawatsch, F. Auger, in *Time-Frequency Analysis*, vol. 36 (Wiley, New York, 2010)
27. S. Mallat, *A Wavelet Tour of Signal Processing*, 3rd edn. (Academic Press, Burlington, 2008)
28. L. Stankovic, M. Dakovic, T. Thayaparan, *Time-Frequency Signal Analysis with Applications* (Artech House, Norwood, 2013)
29. R.A. Carmona, W.L. Hwang, B. Torresani, Characterization of signals by the ridges of their wavelet transforms. IEEE Trans. Sig. Proc. **45**(10), 2586–2590 (1997)
30. R.A. Carmona, W.L. Hwang, B. Torresani, Multiridge detection and time-frequency reconstruction. IEEE Trans. Sig. Proc. **47**(2), 480–492 (1999)
31. I. Daubechies, J. Lu, H. Wu, Synchrosqueezed wavelet transforms: an empirical mode decomposition-like tool. Appl. Comput. Harmon. Anal. **30**(2), 243–261 (2011)
32. N. Delprat, B. Escudie, P. Guillemain, R. Kronland-Martinet, P. Tchamitchian, B. Torrésani, Asymptotic wavelet and Gabor analysis: extraction of instantaneous frequencies. IEEE Trans. Inf. Theory **38**(2), 644–664 (1992)
33. D. Iatsenko, P.V.E. McClintock, A. Stefanovska, On the extraction of instantaneous frequencies from ridges in time-frequency representations of signals. submitted for publication [preprint—arXiv:1310.7276], 2014
34. G. Thakur, E. Brevdo, N.S. Fuckar, H.-T. Wu, The Synchrosqueezing algorithm for time-varying spectral analysis: robustness properties and new paleoclimate applications. Sig. Process. **93**(5), 1079–1094 (2013)
35. Y.-C. Chen, M.-Y. Cheng, H.-T. Wu, Nonparametric and adaptive modeling of dynamic seasonality and trend with heteroscedastic and dependent errors [preprint—arXiv:1210.4672], 2012
36. D. Iatsenko, P.V.E. McClintock, A. Stefanovska, Linear and synchrosqueezed time-frequency representations revisited: Overview, standards of use, reconstruction, resolution, concentration, and algorithms. Digital Signal Processing. (in press), 2015, doi:10.1016/j.dsp.2015.03.004
37. H. Abdi, L.J. Williams, Principal component analysis. Wiley Interdiscip. Rev. Comput. Stat. **2**(4), 433–459 (2010)
38. I.T. Jolliffe, *Principal Component Analysis*, 2nd edn. (Springer, Berlin, 2002)
39. P. Comom, Independent component analysis, a new concept? Sig. Process. **36**(3), 287–314 (1994)
40. A. Hyvärinen, E. Oja, Independent component analysis: algorithms and applications. Neural Networks **13**(4–5), 411–430 (2000)
41. H.-P. Kriegel, P. Kröger, E. Schubert, A. Zimek, A general framework for increasing the robustness of PCA-based correlation clustering algorithms. In *Proceedings SSDBM* (Springer, Berlin, 2008), pp. 418–435
42. B. Schölkopf, A. Smola, K.-R. Müller, Nonlinear component analysis as a kernel eigenvalue problem. Neural Comp. **10**(5), 1299–1319 (1998)
43. M. Hožič, A. Stefanovska, Karhunen - Loève decomposition of peripheral blood flow. Physica A **281**, 587–601 (2000)
44. M.E. Davies, C.J. James, Source separation using single channel ICA. Sig. Proc. **87**(8), 1819–1832 (2007)
45. N.E. Huang, Z. Shen, S.R. Long, M.C. Wu, H.H. Shih, Q. Zheng, N. Yen, C.C. Tung, H.H. Liu, The empirical mode decomposition and the Hilbert spectrum for nonlinear and non-stationary time series analysis. Proc. R. Soc. Lond. A **454**, 903–995 (1998)
46. Z. Wu, N.E. Huang, Ensemble empirical mode decomposition: a noise-assisted data analysis method. Adv. Adapt. Data Anal. **1**(1), 1–41 (2009)

Chapter 2
Linear Time-Frequency Analysis

The identification and quantification of the oscillatory components present in a given signal represents a classic problem of signal processing, and one of the most successful approaches for solving it has been through time-frequency analysis. Thus, it is often useful to study the signal's structure in both time and frequency simultaneously, which can be done by considering a specifically constructed projection of the signal onto the time-frequency plane—its time-frequency representation. Such an approach proves to be especially suitable for signals containing many oscillatory components with time-varying amplitudes and/or frequencies, which is a very common scenario for real-life signals [1, 25, 37, 48, 49].

Consider, for example, the activity of the heart: although beating with frequency of 1 Hz on average, the length of the period between the consecutive beats—and therefore the frequency—is time-dependent; moreover, the corresponding variations of the beat-to-beat interval are of high clinical relevance [36]. These variations and their properties are straightforward to analyse in the time-frequency plane, where time-dependent frequency of the heart can be easily traced [9, 10, 17, 18, 25, 26, 50]. The same applies to many other real-life signals, so the TFRs are now routinely applied in almost every area of science, from image processing and finance to geophysics and the life sciences [1, 2, 25, 32–34, 37, 43, 45, 47, 49, 51, 55].

General aspects of time-frequency analysis, and the properties of the different existing TFRs, have been thoroughly discussed in a number of excellent books [1, 8, 14, 16, 21, 22, 30, 37, 46] and reviews [13, 23, 41, 43, 51]. There exist two main TFR classes: linear (such as the WFT and WT [1, 16, 37]) and quadratic (such as the Wigner-Ville [54, 56], Rihaczek [42] and Choi-Williams [12] distributions, see [8, 22, 23] for a more comprehensive list). Although each of these has its own advantages, only the WFT and WT will be considered in this work. This is because, in contrast to quadratic representations, these types offer an easy and straightforward way of extracting and reconstructing individual components present in the signal (see Chap. 3), which property will be of crucial importance in what follows.

The present chapter presents a thorough discussion of the WFT and WT, their implementation, properties and related issues, providing all the information that is needed to understand them and apply them effectively. Particular attention is paid to

D. Iatsenko, *Nonlinear Mode Decomposition*,
Springer Theses, DOI 10.1007/978-3-319-20016-3_2

the practical aspects, but the necessary mathematical theory is also reviewed. The basic definitions, such as the AM/FM component and the analytic signal, as well as the need for time-frequency analysis, are first discussed in Sect. 2.1. The WFT and WT are then defined in Sect. 2.2, which also reviews their basic properties and illustrates the difference between the two. The resolution properties of the WFT and WT are considered in Sect. 2.3, while Sect. 2.4 discusses various aspects of "real-life" TFR usage, namely: signal preprocessing that should be performed; frequency discretization and how to do it appropriately; boundary distortions due to finite signal length; and limits on the frequencies that can be studied reliably in the WFT and WT. The results of the whole chapter are summarized in Sect. 2.5.

Remark 2.0.1 It should be noted, that many advanced methods based on different TFRs have been developed over the last few decades. Examples are the wavelet bispectral analysis [28, 29], wavelet coherence [20, 33, 35] and phase coherence [3, 5, 44] etc. However, it seems in principle impossible to review appropriately all such techniques, given their large number, so they are not considered here. Rather, the emphasis in the present Chapter is placed on the basic time-frequency representations, which provide the foundation for more sophisticated methods, including nonlinear mode decomposition developed later.

2.1 AM/FM Components and the Analytic Signal

One of the basic notions of time-frequency analysis is the *AM/FM component* (or simply *component*), which is defined as a function of time t of form

$$x(t) = A(t)\cos\phi(t) \quad \big(\forall t:\ A(t) > 0,\, \nu(t) \equiv \phi'(t) > 0\big). \tag{2.1}$$

The time-dependent values $A(t)$, $\phi(t)$ and $\nu(t) \equiv \phi'(t)$ are then called the instantaneous amplitude, phase and frequency of the component (2.1) (for a more detailed discussion of their definitions and related issues see [6, 7, 40]).

Given that the signal is known to be of the form (2.1), the natural question is how to find its associated $A(t)$, $\phi(t)$ and $\nu(t)$. The most convenient way of doing this is the analytic signal approach. However, before considering it, a few additional notions should be introduced. Thus, for an arbitrary function $f(t)$, its Fourier transform (FT), positive and negative frequency parts, time-average and standard deviation will be denoted as $\hat{f}(\xi)$, $f^+(t)$, $f^-(t)$, $\langle f(t)\rangle$ and $\mathrm{std}[f(t)]$, respectively:

$$\begin{gathered}
\hat{f}(\xi) \equiv \int_{-\infty}^{\infty} f(t)e^{-i\xi t}dt \;\Leftrightarrow\; f(t) = \frac{1}{2\pi}\int_{-\infty}^{\infty}\hat{f}(\xi)e^{i\xi t}d\xi = \langle f(t)\rangle + f^+(t) + f^-(t),\\
f^+(t) \equiv \frac{1}{2\pi}\int_{0^+}^{\infty}\hat{f}(\xi)e^{i\xi t}d\xi, \quad f^-(t) \equiv \frac{1}{2\pi}\int_{-\infty}^{0^-}\hat{f}(\xi)e^{i\xi t}d\xi,\\
\langle f(t)\rangle = \frac{\int f(t)dt}{\int dt}, \quad \mathrm{std}[f(t)] \equiv \sqrt{\langle [f(t)]^2\rangle - [\langle f(t)\rangle]^2},
\end{gathered} \tag{2.2}$$

where, here and in what follows, the integrals are taken over $(-\infty, \infty)$ if unspecified (or, in practice, over the full time duration of $f(t)$). A simple example is $f(t) = a + b\cos\nu t = a + be^{i\nu t}/2 + be^{-i\nu t}/2$, for which one has $\hat{f}(\xi)/2\pi = a\delta(\xi) + b\delta(\xi-\nu)/2 + b\delta(\xi+\nu)/2$, $\langle f(t)\rangle = a$, $f^{\pm}(t) = be^{\pm i\nu t}/2$ and $\mathrm{std}[f(t)] = b/\sqrt{2}$. Note, that if $f(t)$ is real, then $f(t) - \langle f(t)\rangle = 2\,\mathrm{Re}\ f^{\pm}(t)$ and $\hat{f}(\xi) = [\hat{f}(-\xi)]^* \Rightarrow f^+(t) = [f^-(t)]^*$, where the star denotes complex conjugation.

For a given signal $s(t)$ (which is always assumed to be real in this work), its doubled positive frequency part is called its *analytic signal* and will be denoted as $s^a(t)$:

$$s^a(t) \equiv 2s^+(t) \quad \big(s(t) = \langle s(t)\rangle + \mathrm{Re}[s^a(t)]\big). \tag{2.3}$$

The analytic signal is complex, so its dynamics can easily be separated into amplitude and phase parts. For signals represented by a single component (2.1), the analytic amplitude and phase $A^a(t)$, $\phi^a(t)$ match closely the true amplitude and phase $A(t)$, $\phi(t)$, thus providing an easy way to estimate them:

$$A(t) \approx A^a(t) \equiv |s^a(t)|, \quad \phi(t) \approx \phi^a(t) \equiv \arg[s^a(t)]. \tag{2.4}$$

The approximate equality (2.4) will be called the analytic approximation, and it can alternatively be formulated as $[A(t)\cos\phi(t)]^+ \approx A(t)e^{i\phi(t)}/2$. As stipulated by the Bedrosian theorem [4], this approximation is exact when the spectrum of $A(t)$ lies lower than the spectrum of $e^{i\phi(t)}$, and there are no intersections between the two. For example, in the case of amplitude modulation only, $s(t) = A(t)\cos(\nu t + \varphi)$, (2.4) gives the exact amplitude and phase if all the spectral content of $A(t)$ lies lower than ν, i.e. $\hat{A}(\xi \geq \nu) = 0$. Usually, however, there is a small discrepancy between the true and analytic amplitude/phase (considered in detail in Sect. 7.1), but it is often very small. Thus, there is arguably still nothing better than the analytic signal approach for amplitude and phase estimation in the case of a single AM/FM component (2.1) [52] (except, maybe, the recently proposed direct quadrature method [24]).

However, real-life signals rarely consist of only one component, and they usually also contain noise. In this case, the analytic signal will represent a mix of the amplitude and phase dynamics of all components contained in the signal (additionally corrupted by noise), so their individual parameters cannot be recovered from it. One should therefore employ more sophisticated techniques, able to distinguish the different components within a single time-series. This can be done by using the TFR-based approaches that will be described below. The main assumption behind them (and time-frequency analysis more generally) is that the signal is represented by a sum of AM/FM components, each of which satisfies the analytic approximation (2.4), plus some noise $\eta(t)$:

$$\begin{aligned} &s(t) = \sum_i x_i(t) + \eta(t) = \sum_i A_i(t)\cos\phi_i(t) + \eta(t), \\ &\forall t, i:\ A_i(t) > 0,\ \phi_i'(t) > 0,\ [A_i(t)\cos\phi_i(t)]^+ \approx A_i(t)e^{i\phi_i(t)}/2. \end{aligned} \tag{2.5}$$

Although signal representation (2.5) is not unique, in practice one aims at the sparsest among such representations, i.e. the one characterized by the smallest number of components $x_i(t)$. It is also important to note, that the most accurate estimates of the components' parameters obtainable using any TFR-based method are the corresponding analytic estimates (2.4), so the (best achievable) goal of such methods is to extract the separate analytic signals $x_i^a(t)$ for each of the chosen $x_i(t)$ in (2.5).

The above discussion was related to problems that arise while considering multicomponent signals in the time domain. To make the story complete, the problems that arise when trying to treat such signals in the frequency domain should also be addressed. In practice, any signal can be represented by its Fourier expansion (discrete FT), i.e. as the sum of tones—monochromatic signals $A_i \cos(\nu_i t + \varphi_i)$—with different constant amplitudes A_i, phase shifts φ_i and frequencies ν_i. The same is obviously true for a single AM/FM component, and its Fourier expansion is directly related to the spectrum of the corresponding amplitude and frequency modulations:

$$
\begin{aligned}
x(t) =& A\big(1+\sum_a r_a \cos(\nu_a t + \varphi_a)\big) \cos\big(\nu t + \varphi + \sum_b r_b \sin(\nu_b t + \varphi_b)\big) \\
&= \Big/ \text{expressing } \cos\phi = (e^{i\phi}+e^{-i\phi})/2,\; e^{ia\sin\phi} = \sum_{n=-\infty}^{\infty} J_n(a) e^{in\phi} \Big/ \\
&= Ae^{i(\nu t+\varphi)}\Big[1+\sum_a \frac{r_a(e^{i(\nu_a t+\varphi_a)}+e^{-i(\nu_a t+\varphi_a)})}{2}\Big]\prod_b \sum_{n_b} J_{n_b}(r_b) e^{i(n_b\nu_b t + n_b\varphi_b)} + c.c. \\
&= \sum_{\{n_b\}=\{n_1,n_2,\dots\}\in\mathbb{Z}} \Big[\widetilde{A}_{\{n_b\}} e^{i\widetilde{\varphi}_{\{n_b\}}} e^{i\widetilde{\nu}_{\{n_b\}}t} + \sum_a \frac{r_a}{2} \widetilde{A}_{\{n_b\}} \big(e^{i(\widetilde{\varphi}_{\{n_b\}}+\varphi_a)} e^{i(\widetilde{\nu}_{\{n_b\}}+\nu_a)t} \\
&\quad + e^{i(\widetilde{\varphi}_{\{n_b\}}-\varphi_a)} e^{i(\widetilde{\nu}_{\{n_b\}}-\nu_a)t}\big)\Big] + c.c.,
\end{aligned}
\tag{2.6}
$$

where

$$
\begin{aligned}
\widetilde{A}_{\{n_b\}} &\equiv A \prod_b J_{n_b}(r_b) = A J_{n_1}(r_1) J_{n_2}(r_2)\dots, \\
\widetilde{\nu}_{\{n_b\}} &\equiv \nu + \sum_b n_b \nu_b, \quad \widetilde{\varphi}_{\{n_b\}} \equiv \varphi + \sum_b n_b \varphi_b,
\end{aligned}
\tag{2.7}
$$

and $J_n(r_b) = (-1)^n J_{-n}(r_b)$ denote Bessel functions of the first kind, while *c.c.* stands for the complex conjugate of the preceding expression. Thus, any component (2.6) can be represented as a sum of tones with frequencies $|\nu + \sum_b n_b \nu_b|$ and $|\nu \pm \nu_a + \sum_b n_b \nu_b|$ (for all possible combinations of $a \in \mathbb{N}$ and $\{n_b\} = \{n_1, n_2, \dots\} \in \mathbb{Z}$). But, obviously, not any sum of tones can be represented as the AM/FM component. What make this possible in (2.6) are the specific relationships between the tones' amplitudes, phases and frequencies, which encode the corresponding amplitude and frequency modulations of the component.

Clearly, the representation of $x(t)$ (2.6) as a single entity is more compact and meaningful than its representation as the sum of tones, which corresponds to its FT. Thus, for a multi-component signal $s(t)$ (2.5), each component is encoded as a set of

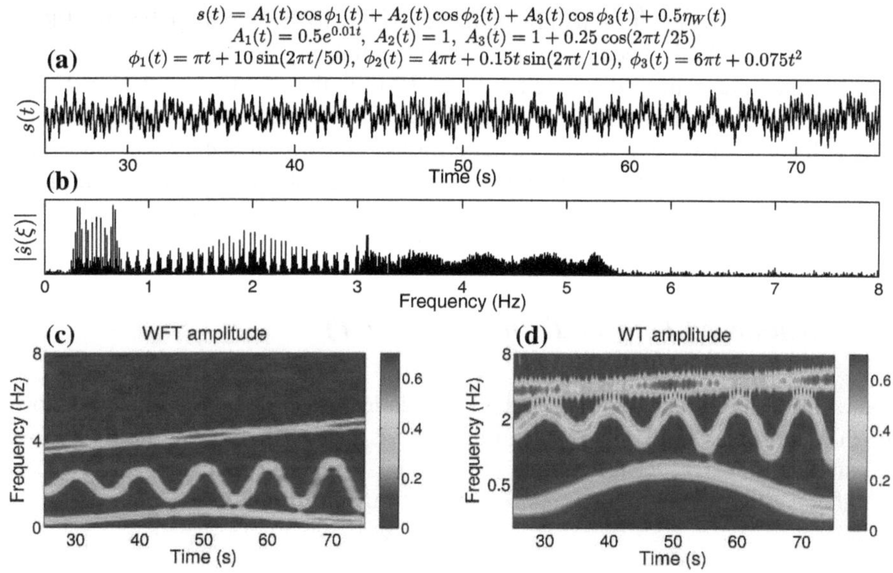

Fig. 2.1 Different representations of a signal composed of three AM/FM components and corrupted by noise $0.5\,\eta_W(t)$, where $\eta_W(t)$ denotes Gaussian white noise of unit deviation; the parameters of the components are described above the *top panel*. **a** Signal in the time domain; **b** signal in the frequency domain, given by its Fourier transform; **c**, **d** signal in the time-frequency domain, given by its WFT and WT (see Sects. 2.2.1 and 2.2.2 below), respectively. Note that for the WT (**d**) the frequency scale is logarithmic

entries in the associated frequency representation $\hat{s}(\xi)$, but it is usually unclear which entries correspond to what component, and which ones are attributable to noise. This is illustrated in Fig. 2.1, which shows an example of signal representation in the time, frequency, and time-frequency domains, with the latter being given by its WFT and WT, to be discussed below. As can be seen, although all representations by definition contain the same amount of information about the signal, in the case of Fig. 2.1 the most readily interpretable view of this information is provided in the time-frequency domain.

2.2 Time-Frequency Representations (TFRs)

As discussed in the previous section, instead of studying a signal in either one of the time ($s(t)$) or frequency ($\hat{s}(\xi)$) domains, it is often more useful to study its properties in time and frequency simultaneously. This can be done by considering specifically constructed projections of the signal onto a time-frequency plane, i.e. TFRs. Such an approach gives the possibility of tracking the evolution of the signal's spectral content in time, which is typically represented by variations of the amplitudes and frequencies of the components from which the signal is composed.

In what follows, attention will be paid exclusively to the two main linear TFRs—the WFT and the WT—because of their special suitability for the extraction of components (the latter will be discussed in more detail in the next Chapter). Note that in the present Sect. the notions of time, frequency and time-frequency resolutions of the TFR will sometimes be used; they will be further clarified in Sect. 2.3 below. There are also many classical books and reviews (e.g. [8, 22, 30, 37]) where an additional discussion of the WFT/WT and their resolution characteristics can be found.

2.2.1 Windowed Fourier Transform (WFT)

The windowed Fourier transform (WFT), also called the short-time Fourier transform or (in a particular form) the Gabor transform, is one of the oldest and thus most-investigated linear TFRs [19, 22, 37]. The WFT $G_s(\omega, t)$ of the signal $s(t)$ can be calculated either in the time domain or in the frequency domain as [27]:

$$G_s(\omega, t) = \int s^+(\tau)g(\tau - t)e^{-i\omega(\tau - t)}d\tau = \frac{1}{2\pi}\int_0^\infty e^{i\xi t}\hat{s}(\xi)\hat{g}(\omega - \xi)d\xi, \quad (2.8)$$

where $g(t)$ is the specified *window function* and $\hat{g}(\xi)$ is its FT. Without loss of generality, the $|\hat{g}(\xi)|$ is considered to have a maximum at $\xi = 0$:

$$\omega_g \equiv \operatorname{argmax} |\hat{g}(\xi)| = 0, \quad (2.9)$$

which ensures that, for a single tone $s(t) = \cos \nu t$, the amplitude of the WFT has a maximum at the frequency $\omega = \nu$ (see (2.10) below). If this is not the case, one should always set ω_g (2.9) to zero manually by considering the demodulated window function $\{g(t), \hat{g}(\xi)\} \rightarrow \{g(t)e^{-i\omega_g t}, \hat{g}(\xi + \omega_g)\}$. An example of the WFT can be seen in Fig. 2.1c.

Remark 2.2.1 Note, that the definition of the WFT (2.8) is slightly different from the conventional form. Thus, only the positive frequency part $s^+(t)$ is taken into account instead of the full $s(t)$, for the reasons that are thoroughly discussed in [27]. Basically, such modification is needed to remove the undesirable interference with negative frequencies (e.g. with $e^{-i\nu t}$ if $s(t) = \cos \nu t$), which might seriously corrupt the representation if it has low resolution in frequency. Nevertheless, in practice this interference is often very small, so that taking the full signal in (2.8) will not usually change anything significantly. Additionally, instead of the traditional $e^{-i\omega\tau}$, which corresponds to "static" phases of the WFT coefficients, there is $e^{-i\omega(\tau - t)}$ in (2.8), which corresponds to "dynamic" phases and is thus more convenient in terms of component reconstruction.

It is easy to see that, for a signal represented by a sum of tones, the corresponding WFT (2.8) takes the form

$$s(t) = \sum_n A_n \cos(\nu_n t + \varphi_n) \Rightarrow G_s(\omega, t) = \frac{1}{2} \sum_n A_n \hat{g}(\omega - \nu_n) e^{i(\nu_n t + \varphi_n)}, \quad (2.10)$$

from which one can gain a first impression of how it works. It should be noted that, although peaks in $|G_s(\omega, t)|$ are all at positive frequencies (as follows from (2.10) and the assumptions made), the WFT (2.8) is defined on the full interval $\omega \in (-\infty, \infty)$, so that one can calculate it for $\omega < 0$ as well (e.g. to trace the peak's tails). Usually, however, this is not needed.

The WFT is an invertible transform, so that the original signal in both time and frequency domains can be recovered from it as (see Sect. 7.3):

$$\begin{aligned} s^a(t) &= C_g^{-1} \int G_s(\omega, t) d\omega, \quad s(t) = \langle s(t) \rangle + \mathrm{Re}[s^a(t)], \\ \hat{s}(\omega > 0) &= \widetilde{C}_g^{-1} \int G_s(\omega, t) e^{-i\omega t} dt, \quad \hat{s}(-\omega) = \hat{s}^*(\omega), \\ C_g &\equiv \frac{1}{2} \int \hat{g}(\xi) d\xi = \pi g(0), \quad \widetilde{C}_g \equiv \int g(t) dt = \hat{g}(0). \end{aligned} \quad (2.11)$$

Among the window functions $g(t)$ that can be used for calculating WFT (2.8) (see Sect. 7.2 for a list of common types and their properties), the most suitable choice seems to be the Gaussian window

$$g(t) = \frac{1}{\sqrt{2\pi} f_0} e^{-t^2/2f_0^2} \Leftrightarrow \hat{g}(\xi) = e^{-f_0^2 \xi^2/2}. \quad (2.12)$$

The time and frequency resolution of the window (2.12) are determined by its spread in time and frequency, which are controlled by its *resolution parameter* f_0. Therefore, increasing f_0 improves the frequency resolution of the resultant WFT, but reduces its time resolution.

The Gaussian window is commonly used on account of its unique property of maximizing the "classic" time-frequency resolution of the transform (to be discussed in Sect. 2.3); the form (2.12), however, seem to have very good resolution properties in general, and not only within the traditional definitions (see Sect. 7.2). In what follows, unless otherwise specified, all simulations are performed using the Gaussian window (2.12) with $f_0 = 1$. Nonetheless, the considerations and formulas of this Chapter apply quite generally for any window function.

In numerical applications, given the signal $s(t_n)$, $n = 1, \ldots, N$ sampled at frequency f_s, its WFT is calculated using the fast Fourier transform (FFT) algorithm, utilizing the frequency domain form of (2.8). The frequency axis is first divided into equidistant bins $\omega_k = (k - k_0)\Delta\omega$, where $\Delta\omega$ is the frequency bin width (its optimal choice and related issues are discussed in Sect. 2.4.2). One then computes the signal's discrete FT $\hat{s}(\xi) = \{\hat{s}(\xi_n)\}$, where $\xi_n/2\pi = (n/N - 1/2) f_s$, $n = 1, \ldots, N$ are the corresponding discrete frequencies, and sets $\hat{s}(\xi_n \leq 0) = 0$. Finally, taking the inverse FT of $\{s(\xi_n)\hat{g}(\omega_k - \xi_n)\}$ at each frequency ω_k gives the full WFT time evolution $G_s(\omega_k, t_{n=1,\dots,N})$ for that frequency bin. A complete numerical implementation, including all relevant issues, is summarized in Sect. 7.5.

2.2.2 *Wavelet Transform (WT)*

The (continuous) wavelet transform (WT) is the other well-known linear TFR which, in contrast to the WFT, has logarithmic frequency resolution [15, 16, 37]; in other respects the two TFRs are quite similar. The WT $W_s(\omega, t)$ of a signal $s(t)$ can be calculated as [27]:

$$W_s(\omega, t) = \int s^+(\tau)\psi^*\Big(\frac{\omega(\tau - t)}{\omega_\psi}\Big)\frac{\omega d\tau}{\omega_\psi} = \frac{1}{2\pi}\int_0^\infty e^{i\xi t}\hat{s}(\xi)\hat{\psi}^*(\omega_\psi\xi/\omega)d\xi, \tag{2.13}$$

where $\psi(t)$ is the chosen *wavelet function*, and

$$\omega_\psi \equiv \text{argmax}\,|\hat{\psi}(\xi)| \tag{2.14}$$

denotes the wavelet peak frequency. Additionally, the wavelet should satisfy the *admissibility condition*

$$\hat{\psi}(0) = \int \psi(t)dt = 0, \tag{2.15}$$

the need for which will become clear below. An example of the WT can be seen in Fig. 2.1d.

Remark 2.2.2 It should be noted, that the definition (2.13) represents a modified version of the traditional one. Thus, the WT is usually defined using the scales $a(\omega)$, while in (2.13) they are already expressed through the frequencies as $a(\omega) = \omega_\psi/\omega$ (this relation establishes that, in the simplest case $s(t) = A\cos(\nu t + \varphi)$, the modulus of the WT $|W_s(\omega, t)|$ will be peaked exactly at the tone frequency $\omega = \nu$). Next, similarly to the WFT (2.8), the WT is calculated using only the positive frequency part of the signal $s^+(t)$ (this is in fact equivalent to using the full signal $s(t)$ but setting $\hat{\psi}(\xi \leq 0) = 0$, though many wavelets, called analytic, have this property from the beginning). The reasoning behind this is the same as for the case of the WFT (see Remark 2.2.1). Finally, there are a few ways of normalizing the WT, and the most commonly used one corresponds to (2.13) multiplied by $\sqrt{\omega_\psi/\omega}$. However, in the latter case the WT amplitude will be biased towards lower ω, amplifying/reducing the peaks associated with each component depending on its frequency. Such a situation does not arise for the normalization (2.13), which therefore seems to be more natural and convenient, especially in terms of the extraction and reconstruction of components discussed in Chap. 3 below. Nevertheless, the two normalizations differ only in terms of convenience and ease of understanding: the correct usage of each will by definition give the same results.

Comparing (2.13) with (2.8), it is clear that the continuous WT can be viewed simply as the WFT with a frequency-varying window, which feature nevertheless gives rise to a different type of resolution. Thus, the analogue of (2.10) for WT (2.13) is

$$s(t) = \sum_n A_n \cos(\nu_n t + \varphi_n) \Rightarrow W_s(\omega, t) = \frac{1}{2} \sum_n A_n \hat{\psi}^*\left(\omega_\psi \frac{\nu_n}{\omega}\right) e^{i(\nu_n t + \varphi_n)}. \quad (2.16)$$

As can be seen, in contrast to the WFT, which takes into account the frequency difference $\nu - \omega$ (2.10), the WT considers the ratio ν/ω (or the difference between logarithms $\log \nu/\omega = \log \nu - \log \omega$), which amounts to the definition of the logarithmic frequency resolution. The natural frequency scale is therefore also logarithmic for the WT, in contrast to the linear scale of the WFT.

The reconstruction formulas in the case of the WT become (see Sect. 7.3)

$$\begin{aligned} s^a(t) &= C_\psi^{-1} \int_0^\infty W_s(\omega, t) \frac{d\omega}{\omega}, \quad s(t) = \langle s(t) \rangle + \mathrm{Re}[s^a(t)], \\ \hat{s}(\omega > 0) &= \widetilde{C}_\psi^{-1} \int W_s(\omega, t) e^{-i\omega t} dt, \quad \hat{s}(-\omega) = \hat{s}^*(\omega), \qquad (2.17) \\ C_\psi &\equiv \frac{1}{2} \int_0^\infty \hat{\psi}^*(\xi) \frac{d\xi}{\xi}, \quad \widetilde{C}_\psi \equiv \left[\int \psi(t) e^{-i\omega_\psi t} dt \right]^* = \hat{\psi}^*(\omega_\psi). \end{aligned}$$

Note that, in contrast to the WFT (2.11), the signal is reconstructed from the WT by integration of $W_s(\omega, t)$ over the logarithmic scale $d\omega/\omega = d \log \omega$, which is standard for the WT-based measures.

From (2.16) and (2.17), the need for the admissibility condition (2.15) now becomes clear. Thus, for wavelets not satisfying (2.17) the value of C_ψ in (2.17) is infinite, and signal reconstruction from the WT becomes impossible. Even more importantly, from (2.16) it follows that, if the wavelet FT does not vanish at zero frequency $\hat{\psi}(0) \neq 0$, then each component will be spread over the whole frequency range of the WT, leading to a highly corrupted representation. For example, for a single tone signal $s(t) = A \cos \nu t$, the WT amplitude approaches $|W_s(\omega, t)| = A|\hat{\psi}(\omega_\psi \nu/\omega)| \to A|\hat{\psi}(0)|$ as $\omega \to \infty$. Hence, $|\hat{\psi}(0)|$ determines the minimum level to which the WT amplitude for each component decays: if it is nonzero then even components with infinitely distant frequencies will interfere with each other. Drawing an analogy with the WFT, use of the wavelet with $\hat{\psi}(0) \neq 0$ corresponds to the use of the window with $\hat{g}(\xi \to \infty) \neq 0$ in (2.8), which is obviously inappropriate. Each wavelet should therefore be admissible, i.e. should satisfy (2.15).

Remark 2.2.3 It is often not recognized that, in contrast to the case of the WFT, rescaling of the wavelet function $\{\psi(t), \hat{\psi}(\xi)\} \to \{\psi(rt), \hat{\psi}(\xi/r)\}$ has no effect on the resultant WT. Thus, what matters is $\hat{\psi}(\omega_\psi \nu/\omega)$, while direct rescaling changes both $\hat{\psi}(\xi)$ and the peak frequency $\omega_\psi \to r\omega_\psi$, having no overall effect when substituted in (2.13) and (2.16). As an illustration, the parameter k in the wavelet $\hat{\psi}(\xi) = \xi^a e^{-k\xi^b}$ is completely redundant, because this wavelet can be represented in the form $\hat{\psi}(\xi) = k^{-a/b}(k^{1/b}\xi)^a e^{-(k^{1/b}\xi)^b}$ which, up to the constant multiplier, is equivalent to $\hat{\psi}(\xi) = \xi^a e^{-\xi^b}$. To really change the time and frequency resolutions of the wavelet, one needs to change either its ω_ψ while trying to preserve the spread and the form of $|\hat{\psi}(\xi)|$, or *vice versa* (see Sect. 2.3.1). Furthermore, these two approaches are completely equivalent, as can be seen from (2.13), so it is redundant to define parameters controlling both the peak frequency of a wavelet and its spread.

Among the wavelet functions $\psi(t)$ used for the continuous WT (2.13) (see Sect. 7.2 for a list of common types and their properties), the most popular choice is the Morlet wavelet [38]

$$\begin{aligned} \psi(u) &= \frac{1}{\sqrt{2\pi}}\left(e^{i2\pi f_0 u} - e^{-(2\pi f_0)^2/2}\right)e^{-u^2/2}, \\ \hat{\psi}(\xi) &= e^{-(\xi-2\pi f_0)^2/2}\left(1 - e^{-2\pi f_0 \xi}\right), \end{aligned} \tag{2.18}$$

where the *admissibility term* $\sim e^{-(2\pi f_0)^2/2}$ is needed to satisfy (2.15). The Morlet wavelet is regarded as the wavelet analogy to the Gaussian window (2.12), and is used so commonly because of the widespread belief that it has the best resolution properties. This is, however, not true. Thus, as will be discussed in Sect. 2.3 below, the wavelet (2.18) does indeed nearly maximize the "classic" time-frequency resolution (provided the admissibility term is small), but this traditional measure was originally devised for the WFT and is completely inappropriate for the WT.

Taking into account the logarithmic frequency resolution of the WT, a more "correct" wavelet analogy of the Gaussian window (2.12) would be the lognormal wavelet

$$\hat{\psi}(\xi > 0) \sim e^{-(2\pi f_0 \log \xi)^2/2}, \quad \omega_\psi = 1, \tag{2.19}$$

where the resolution parameter f_0 has a meaning similar to that for the Gaussian window (2.12), controlling the time and frequency resolutions of the resultant WT.

It turns out (see Sect. 7.2) that the resolution properties of the wavelet (2.19) are generally slightly better than that of the Morlet wavelet (2.18). Apart from this, the lognormal wavelet has a variety of other advantages. Thus, it is "infinitely admissible" (in contrast to the Morlet wavelet), i.e. all its moments $\int \xi^{-n}\hat{\psi}(\xi)d\xi/\xi$ $(n \geq 0)$ are finite; this allows direct reconstruction of any order time-derivatives of the component's amplitude and phase from its WT (see Sects. 3.2 and 7.3 below). Additionally, the wavelet (2.19) is analytically tractable, allowing one to obtain C_ψ (2.17) and many other quantities in the explicit form. Therefore, while the considerations of the present Chapter are applicable to any wavelet type, in what follows, unless otherwise specified,, all the simulations and examples will be based on the lognormal wavelet with $f_0 = 1$.

In numerical applications, given the signal $s(t_n), n = 1, .., N$ sampled at frequency f_s, its WT is calculated using the frequency domain form of (2.13) and taking advantage of the FFT algorithm. The procedure is quite similar to that for the WFT calculation. The frequency axis is first partitioned into equilogspaced bins $\omega_k/2\pi = 2^{(k-k_0)/n_v}$, where the *number-of-voices* n_v determines the fineness of this partition and has the meaning of the number of frequency bins in each diadic interval (its selection is discussed in Sect. 2.4.2). One then computes the signal's discrete FT $\hat{s}(\xi) = \{\hat{s}(\xi_n)\}$, where $\xi_n/2\pi = (n/N - 1/2)f_s$, $n = 1, 2, .., N$ denote the corresponding discrete frequencies, and one sets $\hat{s}(\xi_n \leq 0) = 0$. Finally, taking the inverse Fourier transform of $\{\hat{s}(\xi_n)\hat{\psi}^*(\omega_\psi \xi_n/\omega_k)\}$ at each frequency ω_k gives the full time-evolution of the WT $W_s(\omega_k, t_{n=1,...,N})$ for this frequency bin. A complete numerical implementation, including all relevant issues, is summarized in Sect. 7.5.

2.2.3 Difference Between the WFT and the WT

Figure 2.2 compares the WFT and WT calculated for the same signal. As mentioned before, the main distinction between the two TFR types lies in their different kinds of frequency resolution: linear for the WFT, and logarithmic for the WT. Thus, if the signal contains two tones $\cos \nu_{1,2}t$, then the WFT takes account of their frequency difference $\nu_2 - \nu_1$, whereas what matters for the WT is their frequency ratio ν_2/ν_1 (or difference between logarithms of their frequencies). For example, if there are two tones with frequencies 0.05 and 0.125 Hz, corresponding to periods of 20

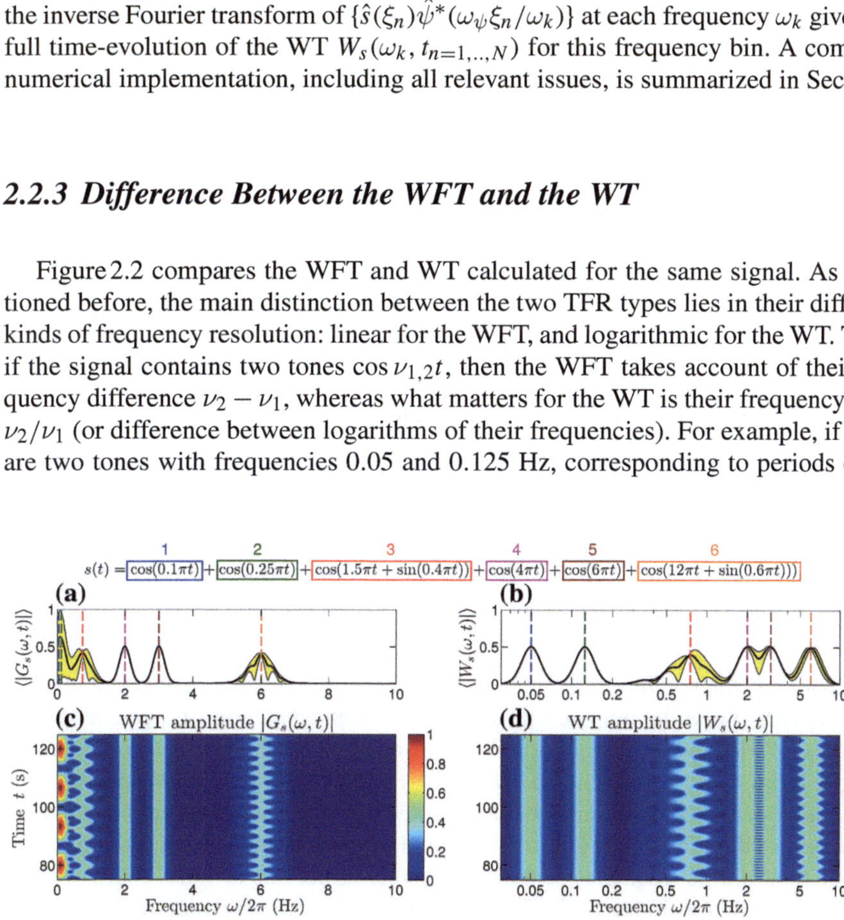

Fig. 2.2 Comparison of the WFT and WT of the same signal consisting of six components, as shown above the figure. **a**, **b** Mean WFT and WT amplitudes, respectively; the *yellow regions* indicate 95 % ranges for the corresponding amplitudes at each frequency; the *vertical dashed lines* show mean frequencies of the components, with their *colors* being linked to the *colors of boxes* at the *top* of the figure. **c**, **d** Full WFT and WT amplitudes, respectively. Note, that for the WT in (**b**, **d**) the frequency scale is logarithmic, in contrast to the linear scale for the WFT in (**a**, **c**). The signal was sampled for 200 s at 50 Hz

and 8 s (components 1,2 in Fig. 2.2), then they will be much better resolved in the WT ($\nu_2/\nu_1 - 1 = 1.5$) than in the WFT ($(\nu_2 - \nu_1)/2\pi = 0.075$). At the same time, if there are two tones with frequencies 2 and 3 Hz, corresponding to periods 0.5 and ≈ 0.33 s (components 4 and 5 in Fig. 2.2), then they will be much better resolved in the WFT ($(\nu_2 - \nu_1)/2\pi = 1$ Hz) than in the WT ($\nu_2/\nu_1 - 1 = 0.5$).

On the other hand, the WFT has the same time resolution at each frequency, while that for the WT increases with frequency, being proportional to ω (see Sect. 2.3 below). This means that, for fixed window/wavelet parameters, the higher the frequency, the more time-variability is allowed for components to be reliably represented in the WT, while the WFT does not discriminate in this respect. For example, if one has a frequency-modulated component $\cos(\nu t + \sin \nu_b t)$, the WFT accounts for the value of ν_b, while the WT considers ν_b/ν. Thus, if one has a component with a mean frequency of 0.75 Hz subject to sinusoidal modulation at 0.2 Hz (component 3 in Fig. 2.2), it will be better represented in the WFT ($\nu_b/2\pi = 0.2$) than in the WT ($\nu_b/\nu \approx 0.27$). On the other hand, if one has a component with mean frequency 6 Hz and frequency modulation at 0.3 Hz (component 6 in Fig. 2.2), it will be represented more reliably in the WT ($\nu_b/\nu = 0.05$) than in the WFT ($\nu_b/2\pi = 0.3$).

Remark 2.2.4 Note, that the assessment of the quality of representation of two tones in the WFT and the WT based on the direct comparison between the corresponding $(\nu_2 - \nu_1)/2\pi$ and $\nu_2/\nu_1 - 1$ is valid only within the assumption that the WFT and the WT have the same frequency resolution properties at $\omega = 2\pi$ (see Sect. 2.3). This is true for the case of Fig. 2.2, where the WFT and WT are calculated using Gaussian window (2.12) and lognormal wavelet (2.19) with the same resolution parameters $f_0 = 1$, respectively. Likewise, in the discussion about representation of the frequency-modulated components it was assumed that the time resolution properties of the WFT and the WT are the same at $\omega = 2\pi$, which is almost true in the present case (see Sect. 2.3 below). The related considerations, however, are only approximately valid because of the difference between the fixed and frequency-dependent time resolutions of the WFT and the WT, respectively.

Obviously, the resolution parameter f_0 (or other window/wavelet parameters, if present) can always be adjusted to improve the representation of any one chosen component. However, there is often no universal f_0 suitable for all components present in the signal, so the choice between the WFT and WT depends on how many such components can in principle be represented reliably. Therefore, contrary to what is sometimes thought, the WT is not in general superior to the WFT: it just considers all on a logarithmic frequency scale, and whether or not this is more useful than the linear frequency resolution of the WFT depends on the signal structure.

Summarizing, the WT is most suitable when the ratios of frequencies of the underlying oscillations and their relative (to frequency) modulations are more consistent with each other than the corresponding frequency differences and absolute amplitude/frequency variations. In other words, the WT is preferred to the WFT when the lower-frequency components are less time-varying and closer in frequency than the components at higher frequencies. Otherwise, the WFT appears to be more appropriate (see also the related discussion and additional examples on pp. 126–134 of

[37]). Note, however, that many real signals have structure that is more suited to studies based on the WT. Additionally, one can often analyse a variety of time-series of different kinds using the same wavelet parameters, while for the WFT the window parameters should be adjusted for each particular case. Usually, the WT is also computationally cheaper due to the logarithmic frequency scale, requiring fewer bins to cover the same frequency range. On the other hand, the WFT generally has better resolution properties (see Sect. 7.2).

2.3 Time-, Frequency- and Time-Frequency Resolution

The notions of time, frequency and time-frequency resolutions have been often used in the preceding section, but were defined only briefly. In this section different resolution characteristics are considered in detail, and the definitions are made more precise.

2.3.1 General Formulation

Consider a signal $s_{\nu\nu}(t)$ ($s_{\tau\tau}(t)$) consisting of two frequency events—tones (time events—delta-peaks), so that its WFT (2.8) and WT (2.13) are

$$\begin{aligned} &s_{\nu\nu}(t) = \cos(\nu t) + \cos((\nu + \Delta\nu)t + \Delta\varphi) \\ &\Rightarrow \begin{cases} G_s(\omega, t) = \frac{1}{2}[\hat{g}(\omega - \nu) + \hat{g}(\omega - \nu - \Delta\nu)e^{i\Delta\nu t}e^{i\Delta\varphi}]e^{i\nu t}, \\ W_s(\omega, t) = \frac{1}{2}[\hat{\psi}^*(\omega_\psi \nu/\omega) + \hat{\psi}^*(\omega_\psi(\nu + \Delta\nu)/\omega)e^{i\Delta\nu t}e^{i\Delta\varphi}]e^{i\nu t}, \end{cases} \end{aligned} \tag{2.20}$$

$$\begin{aligned} &s_{\tau\tau}(t) = \delta(t - \tau) + \delta(t - \tau - \Delta\tau)e^{i\Delta\varphi} \\ &\Rightarrow \begin{cases} G_s(\omega, t) \cong [g(\tau - t) + g(\tau + \Delta\tau - t)e^{-i\omega\Delta\tau}e^{i\Delta\varphi}]e^{i\omega(t-\tau)}, \\ W_s(\omega, t) \cong \frac{\omega}{\omega_\psi}[\psi^*(\omega(\tau - t)/\omega_\psi) + \psi^*(\omega(\tau + \Delta\tau - t)/\omega_\psi)e^{i\Delta\varphi}]. \end{cases} \end{aligned} \tag{2.21}$$

Everywhere in this section, e.g. in (2.21), the symbol "$\cong$" denotes equality up to an error associated with the difference between the WFT/WT (2.8), (2.13) and its form as calculated using the full signal $s(t)$ instead of the corresponding positive frequency part $s^+(t)$; see [27] for a detailed discussion of this issue and the quality of the approximation in (2.21). Note that, if redefining WFT/WT to use the full signal, the approximate equality will migrate from $s_{\tau\tau}(t)$ (2.21) to $s_{\nu\nu}(t)$ (2.20) [27].

Remark 2.3.1 Although in this work the original signal is assumed to be real, $s_{\tau\tau}(t)$ (2.20) represents an exception. This is because, without the phase shift introduced by $e^{i\Delta\phi}$, the analogy between $s_{\tau\tau}(t)$ and $s_{\nu\nu}(t)$ would be incomplete while, intuitively, the time and frequency domains should have equal rights. In fact, since the WFT (2.8) and WT (2.13) are based on the positive frequency part of the signal, the resultant TFR will remain the same if instead of $s_{\nu\nu}(t)$ (2.20) one considers $s^{+}_{\nu\nu}(t) = [e^{i\nu t} + e^{i(\nu+\Delta\nu)t}e^{i\Delta\varphi}]/2$. The latter has the Fourier transform $\hat{s}^{+}_{\nu\nu}(\xi) = [\delta(\xi-\nu) + \delta(\xi-\nu-\Delta\nu)e^{i\Delta\varphi}]/2$, which is fully analogous to $s_{\tau\tau}(t)$. Nevertheless, it turns out that in all the following considerations nothing changes if one restricts $s_{\tau\tau}(t)$ to be real, which is satisfied for $\Delta\varphi = 0, \pi$ in (2.21). For example, in Sect. 2.3.4 below, averaging over these two values ($\Delta\varphi = 0, \pi$) in the corresponding reconstruction errors will give the same result as averaging over the full range $\Delta\varphi \in [0, 2\pi]$.

It seems reasonable to define the time (frequency) resolution γ_t (γ_ω) of the transform as the reciprocal of the minimum time (frequency) difference $\Delta\tau$ ($\Delta\nu$) in $s_{\tau\tau}(t)$ ($s_{\nu\nu}(t)$) for which two delta-peaks (tones) can still be reliably resolved in the TFR:

$$\gamma_t(\nu,\tau) = 1/\Delta\tau_{\min}(\nu,\tau), \quad \gamma_\omega(\nu,\tau) = 1/\Delta\nu_{\min}(\nu,\tau). \tag{2.22}$$

However, the meaning of "reliably resolved" still remains imprecise, and will be dealt with later. Note, that the definitions (2.22) represent the most general case, where the resolutions and minimal differences are "localized", i.e. allowed to depend on both time and frequency. This is not only useful for the WT, which is characterized by frequency-dependent resolution properties, but also allows one to consider more complicated cases, e.g. the WFT with a time-dependent resolution parameter $f_0 \to f_0(t)$.

The joint time-frequency resolution $\gamma_{\omega t}$ is most often defined as the reciprocal of the area of the minimal resolvable square $[\nu, \nu+\Delta\nu_{\min}(\nu,\tau)] \times [\tau, \tau+\Delta\tau_{\min}(\nu,t)]$, being equal to the product of the time and frequency resolutions. However, within such a square $\gamma_t(\nu,\tau)$ and $\gamma_\omega(\nu,\tau)$ can vary considerably, so that these variations should be taken into account to make $\gamma_{\omega t}$ more meaningful. The latter can therefore be defined as

$$\gamma_{\omega t}(\nu,\tau) = \left[\frac{\int_\nu^{\nu+\Delta\nu_{\min}(\nu,\tau)} \Delta\tau_{\min}(\omega,\tau)d\omega \int_\tau^{\tau+\Delta\tau_{\min}(\nu,\tau)} \Delta\nu_{\min}(\nu,t)dt}{\Delta\tau_{\min}(\nu,\tau)\Delta\nu_{\min}(\nu,\tau)}\right]^{-1}. \tag{2.23}$$

Note that, if neither $\Delta\nu_{\min}$ nor $\Delta\tau_{\min}$ depends on time or frequency, then one has the traditional $\gamma_{\omega t} = \gamma_\omega\gamma_t$.

The definitions (2.22) and (2.23) remain valid for any signal representation, not only the WFT and WT. For example, the time domain representation $s(t)$ has infinite time resolution, since, theoretically, two delta-peaks can be resolved for any $\Delta\tau$ in $s_{\tau\tau}(t)$, but zero frequency resolution, as both tones in $s_{\nu\nu}(t)$ have nonzero entries at almost all times, and thus cannot in principle be resolved, no matter how large $\Delta\nu$ is. On the other hand, the frequency domain representation (Fourier transform) $\hat{s}(\xi)$ has infinite frequency resolution, since (in theory) two tones can be perfectly resolved in $\hat{s}_{\nu\nu}(\xi)$ for however small frequency difference $\Delta\nu$; but, at the same time, it has zero time resolution, because the two delta-peaks are spread over all frequencies in $\hat{s}_{\tau\tau}(\xi)$.

TFRs, on the other hand, represent a "mix" of the time domain and frequency domain representations. As a result, both time and frequency resolutions are finite for them. This can clearly be seen from the WFT and WT for the two-peak and two-tone signals (2.21), (2.20). Thus, since the window/wavelet function has non-zero supports in both time and frequency, delta-peaks in $s_{TT}(t)$ (tones in $s_{\nu\nu}(t)$) will interfere in the WFT/WT for small enough $\Delta\tau$ ($\Delta\nu$), making their accurate separation and reconstruction impossible (e.g. see the components 1, 2 in Fig. 2.2a).

From (2.20) and (2.21) one can see that the interference between two tones or between two delta-peaks in the WFT does not depend on ν or τ. Therefore, for any meaningful definition of $\Delta\tau_{\min}$ and $\Delta\nu_{\min}$, they also should not depend on time or frequency. Furthermore, it is also clear that rescaling of the WFT window function $g(t) \to g(t/r) \Leftrightarrow \hat{g}(\xi) \to \hat{g}(r\xi)$ changes its resolution properties as $\{\Delta\tau_{\min}, \Delta\nu_{\min}\} \to \{r\Delta\tau_{\min}, \Delta\nu_{\min}/r\}$. This reflects the repeatedly-mentioned trade-off between time and frequency resolutions, being a manifestation of the time-frequency uncertainty principle, which excludes the possibility of simultaneous sharp localization in time and frequency [30, 37]. Thus, without changing the form of the window function, there is only the possibility of rescaling both resolutions, increasing one and decreasing the other, but not treating them separately. Their product, which in the case of the WFT is equal to the joint time-frequency resolution $\gamma_{\omega t}$ (2.23), remains fixed under such rescaling, thus representing an important characteristic of the window function: the higher it is, the better the trade-off that is possible.

Using the same arguments as for the WFT, from (2.21) and (2.20) it follows that for the WT both $\Delta\tau_{\min}$ and $\Delta\nu_{\min}$ depend on frequency, but not time, so that $\gamma_t = \gamma_t(\nu)$, $\gamma_\omega = \gamma_\omega(\nu)$. Next, one also has $\Delta\tau_{\min}(r\nu) = \Delta\tau_{\min}(\nu)/r$ and $\Delta\nu_{\min}(r\nu) = r\Delta\nu_{\min}(\nu)$, so that the time resolution of the WT increases with frequency, while the frequency resolution decreases. Nevertheless, as will be seen, the time-frequency resolution (2.23) does not depend on ν or τ, being fixed for the specified wavelet parameters.

As mentioned above (see Remark 2.2.3), the direct rescaling $\{\psi(t), \hat{\psi}(\xi)\} \to \{\psi(t/r), \hat{\psi}(r\xi)\}$ does not change anything for the WT, and to tune the resolutions at a particular frequency one needs to change the wavelet peak frequency $\omega_\psi \to r\omega_\psi$ while preserving the spreads and the forms of $\{\psi(t), \hat{\psi}(\xi)\}$ (or vice versa). If such a procedure was possible, it would lead to the same trade-off rule as for the WFT, i.e. $\gamma_t(\nu) \to \gamma_t(\nu)/r$, $\gamma_\omega(\nu) \to r\gamma_\omega(\nu)$, with $\gamma_{\omega t}$ remaining fixed. However, given restrictions such as the admissibility condition (2.15), changing the wavelet peak frequency will inevitably affect its form and/or its spread (on the linear scale). Thus, e.g. for the Morlet wavelet (2.18) a change of ω_ψ (by varying f_0) will be accompanied by a simultaneous change of its form because of the admissibility term; for the lognormal wavelet ω_ψ is fixed, while changing f_0 will change wavelet spread in frequency on the logarithmic scale, but on the linear scale $\hat{\psi}(\xi)$ will become more asymmetric, affecting the time domain form $\psi(t)$. Because of these issues, there does not seem to be a way of changing the resolution properties of the WT without altering its time-frequency resolution, in contrast to the WFT. Furthermore, $\gamma_{\omega t}$ (whichever way defined) progressively worsens with an increase of the wavelet time resolution (decrease of f_0), and it seems in principle impossible to reach a very sharp time-localization in the WT (see Sect. 7.2).

2.3.2 Classical Definitions and Their Flaws

Although $\Delta\tau_{\min}$ and $\Delta\nu_{\min}$ in (2.22) were defined respectively as the minimum time and frequency difference which can reliably be resolved in the TFR, the meaning of "reliably resolved" remains mathematically unclear. Based on how it is defined, one can characterize the resolution properties of the TFR in different ways.

The traditional approach [30, 37] implicitly assumes two delta-peaks (tones) to be well-resolved if the time (frequency) distance between them exceeds some number of standard deviations of the squared window/wavelet function in the time (frequency) domain. Within this framework, for the WFT one has

$$\begin{aligned}
&\Delta\nu_{\min}^{(cl)} = k_1\Delta_\omega, \quad \Delta\tau_{\min}^{(cl)} = k_2\Delta_t,\\
&\Delta_\omega^2 = E_g^{-1}\frac{1}{2\pi}\int(\omega-\omega_c)^2|\hat{g}(\omega)|^2d\omega, \quad \omega_c \equiv E_g^{-1}\frac{1}{2\pi}\int\omega|\hat{g}(\omega)|^2d\omega,\\
&\Delta_t^2 = E_g^{-1}\int(t-t_c)^2|g(t)|^2dt, \qquad t_c \equiv E_g^{-1}\int t|g(t)|^2dt,\\
&E_g \equiv \frac{1}{2\pi}\int|\hat{g}(\xi)|^2d\xi = \int|g(t)|^2dt \quad \text{(by Parseval's identity)},
\end{aligned} \tag{2.24}$$

where $k_{1,2}$ are implicitly assumed to be the same for all window functions. Obviously, such a definition is far from universal, since for different windows different number of standard deviations are needed to resolve the two tones/delta-peaks. As an illustrative example, the window with asymptotics $|\hat{g}(\xi\to\pm\infty)| \sim |\xi|^{-1.25}$ has infinite variance of $|\hat{g}(\xi)|^2$, but can still be used and allows for an accurate resolution and reconstruction of the two tones for high enough frequency difference between them.

For the WT, the classic variance-based framework takes the form

$$\begin{aligned}
&\Delta\nu_{\min}^{(cl)}(\nu) = k_1\frac{\nu}{\omega_\psi}\Delta_\omega, \quad \Delta\tau_{\min}^{(cl)}(\nu) = k_2\frac{\omega_\psi}{\nu}\Delta_t,\\
&\Delta_\omega^2 = E_\psi^{-1}\frac{1}{2\pi}\int(\omega-\omega_c)^2|\hat{\psi}(\omega)|^2d\omega, \quad \omega_c \equiv E_\psi^{-1}\frac{1}{2\pi}\int\omega|\hat{\psi}(\omega)|^2d\omega,\\
&\Delta_t^2 = E_\psi^{-1}\int(t-t_c)^2|\psi(t)|^2dt, \qquad t_c \equiv E_\psi^{-1}\int t|\psi(t)|^2dt,\\
&E_\psi \equiv \frac{1}{2\pi}\int|\hat{\psi}(\xi)|^2d\xi = \int|\psi(t)|^2dt \quad \text{(by Parseval's identity)},
\end{aligned} \tag{2.25}$$

where $k_{1,2}$ are again implicitly assumed to be the same for all wavelet functions. This approach has the same drawbacks as (2.24) for the WFT. However, in the case of the WT it is actually not appropriate at all, at least in terms of the frequency resolution. Thus, as can be seen from (2.16), the tones are represented in the WT as terms $\sim \hat{\psi}(\omega_\psi\nu/\omega)$, so that the decay of their contribution as $\omega\to\infty$, determined by the behavior of $\hat{\psi}(\xi)$ as $\xi\to 0$, will obviously have a big effect on the frequency resolution. At the same time, the usual variance Δ_ω^2 (2.25) takes no account of this fact, e.g. being invariant under $\hat{\psi}(\xi) \to \hat{\psi}(\xi + \omega_\psi)$, that makes the wavelet inadmissible (in which case tones that are infinitely distant in frequency still interfere, so that the frequency resolution becomes effectively zero). Therefore, for wavelets, it seems more appropriate to study at least the variance of $|\hat{\psi}(\omega_\psi/\xi)|^2$, but by no means that of $|\hat{\psi}(\xi)|^2$.

For both WFT and WT, the classic time-frequency resolution measure is taken as $\gamma_{\omega t}^{(cl)} = [\Delta_\omega \Delta_t]^{-1}$, with Δ_ω, Δ_t being given by (2.24) for the WFT and by (2.25) for the WT (note also the difference between $\gamma_{\omega t}^{(cl)}$ and (2.23) for the latter). It can be shown [1, 30, 37], that this measure attains its maximum for the Gaussian window (2.12) and (up to the effect of the admissibility term $\sim e^{-(2\pi f_0)^2/2}$) for the Morlet wavelet (2.18). However, as follows from the discussion above, only in the case of the WFT does the classic $\gamma_{\omega t}^{(cl)}$ make some sense, though even in this case it remains highly non-universal.

2.3.3 Notion of the Window/Wavelet ϵ-Support

Before proceeding to a reconsideration of the classic definitions, it is useful to introduce the notions of the window/wavelet ϵ-supports in frequency $[\xi_1(\epsilon), \xi_2(\epsilon)]$ and time $[\tau_1(\epsilon), \tau_2(\epsilon)]$, which will be used frequently in what follows. These ϵ-supports are defined as the widest intervals containing the $(1-\epsilon)$ part of the total integrals of the window/wavelet function which appear in $C_{g,\psi}$ and $\widetilde{C}_{g,\psi}$ (2.11), (2.17). As will be seen below, they are directly related to the accuracy with which the components can be recovered from the WFT/WT, and thus can be used effectively for quantifying it.

Considering first the WFT, for an arbitrary window function, including functions that are not always positive and can be oscillating or complex, the corresponding definitions are

$$
\begin{aligned}
R_g(\omega) &\equiv \frac{\int_{-\infty}^{\omega} \hat{g}(\xi)d\xi}{\int \hat{g}(\xi)d\xi} = C_g^{-1}\frac{1}{2}\int_{-\infty}^{\omega} \hat{g}(\xi)d\xi,\\
\xi_{1,2}(\epsilon) &: \ |R_g(\xi \le \xi_1)| \le \epsilon/2, \ |1 - R_g(\xi \ge \xi_2)| \le \epsilon/2,\\
P_g(\tau) &\equiv \frac{\int_{-\infty}^{\tau} g(t)dt}{\int g(t)dt} = \widetilde{C}_g^{-1}\int_{-\infty}^{\tau} g(t)dt,\\
\tau_{1,2}(\epsilon) &: \ |P_g(\tau \le \tau_1)| \le \epsilon/2, \ |1 - P_g(\tau \ge \tau_2)| \le \epsilon/2.
\end{aligned}
\tag{2.26}
$$

Evidently, $|R_g(\omega)|$ and $|1 - R_g(\omega)|$ quantify the relative parts of $\hat{g}(\xi)$ that are contained in the ranges $\xi < \omega$ and $\xi > \omega$, respectively, while the values $\xi_{1,2}(\epsilon)$ specify the limits within which the $(1-\epsilon)$ part of the window FT resides. In the same manner, $|P_g(\tau)|$ and $|1 - P_g(\tau)|$ reflect the relative parts of $g(t)$ contained in the ranges $t < \tau$ and $t > \tau$, respectively, while $[\tau_1(\epsilon), \tau_2(\epsilon)]$ represents the region encompassing its $(1-\epsilon)$ part. The inequalities in the definitions of $\xi_{1,2}(\epsilon)$ $(\tau_{1,2}(\epsilon))$ are needed only if $\hat{g}(\xi)$ ($g(t)$) is not strictly positive, or complex, to ensure that the integral of the latter over any frequency (time) region containing the ϵ-support $[\xi_1(\epsilon), \xi_2(\epsilon)]$ ($[\tau_1(\epsilon), \tau_2(\epsilon)]$) will always approximate the corresponding full integral with relative error not higher than ϵ.

Considering the single tone $s(t) = \cos \nu t \Rightarrow G_s(\omega, t) = \hat{g}(\omega - \nu)e^{i\nu t}/2$, it is clear that its WFT at frequencies $\omega < \xi$ ($\omega > \xi$) will contain the $|R_g(\xi - \nu)|$ ($|1 - R_g(\xi - \nu)|$) part of the signal. Furthermore, its $(1 - \epsilon)$ part will be contained in the frequency range $[\nu + \xi_1(\epsilon), \nu + \xi_2(\epsilon)]$, so that e.g. for real $\hat{g}(\xi)$ the corresponding signal reconstructed by (2.11) from the WFT in this range will be $(1 - \epsilon)\cos \nu t$.

Likewise, for the delta-peak $s(t) = \delta(t - \tau) \Rightarrow G_s(\omega, t) \cong g(\tau - t)e^{i\omega(t-\tau)}$ the WFT at times $t < t_0$ ($t > t_0$) will contain the $\cong |1 - P_g(\tau - t_0)|$ ($\cong |P_g(\tau - t_0)|$) part of the signal, while its $(1 - \epsilon)$ part will be contained in the time interval $\cong [\tau - \tau_2(\epsilon), \tau - \tau_1(\epsilon)]$. Thus, e.g. for the real $g(t)$ the delta-peak's FT reconstructed by (2.11) from this interval will be $\hat{s}(\xi) \cong (1 - \epsilon)e^{-i\xi\tau}$.

Similarly to the case of the WFT, the ϵ-supports for the WT are defined based on (2.17) as

$$\begin{aligned}
&R_\psi(\omega) \equiv \frac{\int_0^\omega \hat{\psi}^*(\xi)\frac{d\xi}{\xi}}{\int_0^\infty \hat{\psi}^*(\xi)\frac{d\xi}{\xi}} = C_\psi^{-1}\frac{1}{2}\int_0^\omega \hat{\psi}^*(\xi)\frac{d\xi}{\xi},\\
&\xi_{1,2}(\epsilon):\ |R_\psi(\xi \le \xi_1)| \le \epsilon/2,\ |1 - R_\psi(\xi \ge \xi_2)| \le \epsilon/2,\\
&P_\psi(\tau) \equiv \frac{\int_{-\infty}^\tau \psi^*(t)e^{i\omega_\psi t}dt}{\int \psi^*(t)e^{i\omega_\psi t}dt} = \widetilde{C}_\psi^{-1}\int_{-\infty}^\tau \psi^*(t)e^{i\omega_\psi t}dt,\\
&\tau_{1,2}(\epsilon):\ |P_\psi(\tau \le \tau_1)| \le \epsilon/2,\ |1 - P\psi(\tau \ge \tau_2)| \le \epsilon/2.
\end{aligned} \tag{2.27}$$

Like $|P_g(\tau)|$ in (2.27), $|P_\psi(\tau)|$ ($|1 - P_\psi(\tau)|$) quantifies the relative part of $\psi(t)e^{-i\omega_\psi t}$ contained at $t < \tau$ ($t > \tau$), with $[\tau_1(\epsilon), \tau_2(\epsilon)]$ specifying the interval encompassing its $(1 - \epsilon)$ part. In the same manner, $|R_\psi(\omega)|$ and $\xi_{1,2}(\epsilon)$ are related to the relative part of $\hat{\psi}(\xi)$, taken on a logarithmic scale.

However, due to the scaling nature of the WT, the relationships of (2.27) to real quantities differ slightly from the case of the WFT. Thus, for the single tone $s(t) = \cos(\nu t) \Rightarrow W_s(\omega, t) = \hat{\psi}^*(\omega_\psi \nu/\omega)e^{i\nu t}/2$ the WT at frequencies $\omega < \xi$ ($\omega > \xi$) will contain the $|1 - R_\psi(\omega_\psi \nu/\xi)|$ ($|R_\psi(\omega_\psi \nu/\xi)|$) part of the signal, while its $(1 - \epsilon)$ part will lie in the band $[\omega_\psi \nu/\xi_2(\epsilon), \omega_\psi \nu/\xi_1(\epsilon)]$.

For the delta-function $s(t) = \delta(t - \tau) \Rightarrow W_s(\omega, t) \cong \frac{\omega}{\omega_\psi}\psi^*\left(\frac{\omega(\tau - t)}{\omega_\psi}\right)$ the WT spread in time will vary for different ω. At each frequency the part of the delta-function's total FT contained in the WT at $t < t_0$ ($t > t_0$) will be $\cong |1 - P_\psi(\omega(\tau - t_0)/\omega_\psi)|$ ($\cong |P_\psi(\omega(\tau - t_0)/\omega_\psi)|$), while its $(1 - \epsilon)$ part will reside in the interval $\cong [\tau - \omega_\psi \tau_2(\epsilon)/\omega, \tau - \omega_\psi \tau_1(\epsilon)/\omega]$.

The quantities (2.26), (2.27) are very convenient and will be used extensively below, not only in the present section. For simplicity, $\tau_{1,2}(\epsilon)$ and $\xi_{1,2}(\epsilon)$ denote the respective ϵ-supports both for the window function in the WFT and for the wavelet function in the WT. The meaning will always be clear from the context. Note that the full supports of the window/wavelet in time ($g(t)$, $\psi(t)$) and frequency ($\hat{g}(\xi)$, $\hat{\psi}(\xi)$), whether finite or not, are $[\tau_1(0), \tau_2(0)]$ and $[\xi_1(0), \xi_2(0)]$, respectively.

2.3.4 Reconsidered Definitions

A more universal and appropriate approach (than the traditional variance-based one) is to regard two components as being reliably resolved if they can each be accurately identified and reconstructed from the signal's TFR (i.e. can be recovered with a relative error not exceeding some threshold). Consider the WFT of the two-tone signal (2.20), from which one wants to find the individual analytic signals $x^a_{\nu\nu;1}(t) = e^{i\nu t}$, $x^a_{\nu\nu;2}(t) = e^{i(\nu+\Delta\nu)t}e^{i\Delta\varphi}$ for each of the two tones. At any time t, this can be done by first dividing the frequency range at some $\omega = \omega_x(t)$ into two parts, each responsible for a separate tone, and then integrating the WFT over the corresponding frequency ranges in the same way as in (2.11). This will give the reconstructed analytic signals $\widetilde{x}^a_{\nu\nu;1,2}(t)$ which, using (2.20) and (2.26), can be represented as

$$\begin{aligned}
\widetilde{x}^a_{\nu\nu;1}(t) &= \int_{-\infty}^{\omega_x(t)} G_s(\omega,t)d\omega \\
&= \frac{C_g^{-1}}{2}e^{i\nu t}\left[\int_{-\infty}^{\omega_x(t)} \hat{g}(\omega-\nu)d\omega + e^{i(\Delta\nu t+\Delta\varphi)}\int_{-\infty}^{\omega_x(t)} \hat{g}(\omega-\nu-\Delta\nu)d\omega\right] \\
&= e^{i\nu t}\left[(1-R_g(\nu-\omega_x(t))) + R_g(\omega_x(t)-\nu-\Delta\nu)e^{i(\Delta\nu t+\Delta\varphi)}\right], \\
\widetilde{x}^a_{\nu\nu;2}(t) &= \int_{\omega_x(t)}^{\infty} G_s(\omega,t)d\omega \\
&= \frac{C_g^{-1}}{2}e^{i\nu t}\left[\int_{\omega_x(t)}^{\infty} \hat{g}(\omega-\nu)d\omega + e^{i(\Delta\nu t+\Delta\varphi)}\int_{\omega_x(t)}^{\infty} \hat{g}(\omega-\nu-\Delta\nu)d\omega\right] \\
&= e^{i\nu t}\left[R_g(\nu-\omega_x(t)) + (1-R_g(\omega_x(t)-\nu-\Delta\nu))e^{i(\Delta\nu t+\Delta\varphi)}\right],
\end{aligned} \tag{2.28}$$

where $R_g(x) = 1 - R_g(-x)$ is defined in (2.26).

Obviously, the reconstruction errors $x^a_{\nu\nu;1,2}(t) - \widetilde{x}^a_{\nu\nu;1,2}(t)$ generally depend on the phase-shift $\Delta\varphi$. Therefore, in the corresponding expressions one should take the average over $\Delta\varphi$, which will be denoted as $\langle ... \rangle_{\Delta\varphi}$. The relative errors of each tone's reconstruction $\varepsilon_{\nu\nu;1,2}(\nu,t,\Delta\nu)$ then become

$$\begin{aligned}
\varepsilon^2_{\nu\nu;1,2}(\nu,t,\Delta\nu) &\equiv \frac{\langle |x^a_{\nu\nu;1,2}(t) - \widetilde{x}^a_{\nu\nu;1,2}(t)|^2\rangle_{\Delta\varphi}}{\langle |x^a_{\nu\nu;1,2}(t)|^2\rangle_{\Delta\varphi}} \\
&= |R_g(\nu-\omega_x(t))|^2 + |R_g(\omega_x(t)-\nu-\Delta\nu)|^2.
\end{aligned} \tag{2.29}$$

Note that, in the present case, averaging over $\Delta\varphi$ and time-averaging will give the same results; however, in general the TFR resolution properties can depend on time, and taking the mean over phase-shifts allows one to localize these errors at each t.

The minimum resolvable frequency difference $\Delta\nu_{\min}(\nu, t)$ can be defined as the minimum $\Delta\nu$ in (2.29) for which the total error $\varepsilon_{\nu\nu}(\nu, t, \Delta\nu)$ is still smaller than some accuracy threshold ϵ_r:

$$\Delta\nu \geq \Delta\nu_{\min}(\nu, t) : \; \varepsilon_{\nu\nu}(\nu, t, \Delta\nu) \equiv [\varepsilon^2_{\nu\nu;1}(\nu, t, \Delta\nu) + \varepsilon^2_{\nu\nu;2}(\nu, t, \Delta\nu)]^{1/2} \leq \epsilon_r. \tag{2.30}$$

It can be expressed through the ϵ_r-support of the window in frequency (2.26). Thus, consider the WFT with real, positive and symmetric $\hat{g}(\omega)$, e.g. a Gaussian (2.12). Then it follows from (2.20) that the minimum WFT amplitude between the peaks corresponding to two tones will always appear at $\omega = \nu + \Delta\nu/2$ (unless these two peaks are merged into a single one at some times, which might happen if $\Delta\nu$ is too small). Therefore, in practice the respective frequency regions of the tones will be separated exactly at their average frequency (see Sect. 3.2.2 below), so that one should use $\omega_{\mathrm{x}}(t) = \nu + \Delta\nu/2$ when estimating the errors (2.29); it can also be shown that, in the present case, these errors are minimized by such a choice of $\omega_{\mathrm{x}}(t)$. The overall reconstruction error is then

$$\varepsilon_{\nu\nu}(\Delta\nu) = \left[2|R_g(-\Delta\nu/2)|^2 + 2|R_g(-\Delta\nu/2)|^2\right]^{1/2} = 2|R_g(-\Delta\nu/2)|, \tag{2.31}$$

and, taking into account that $\xi_1(\epsilon) = -\xi_2(\epsilon)$ due to the assumed window symmetry, it follows from (2.30) and (2.26) that the frequency difference for which two tones are recovered with inaccuracy ϵ is exactly equal to the ϵ-support of $\hat{g}(\xi)$.

For other window forms (e.g. asymmetric $\hat{g}(\xi)$) the situation becomes more complicated, but one can still expect to get an overall error of around ϵ when $\Delta\nu = \xi_2(\epsilon) - \xi_1(\epsilon)$. Note that the above considerations hold for reasonably small ϵ, so that $\Delta\nu$ is high enough and there are always two distinct peaks in the WFT amplitude; otherwise, if the peaks are merged at certain times, the actual reconstruction errors will be larger than (2.29).

The case of two delta-peaks (2.21) is closely similar to that of two tones, so the same considerations apply, with just $\xi_{1,2}(\epsilon) \rightarrow \tau_{1,2}(\epsilon)$. Hence, setting ϵ_r as the maximum allowable reconstruction error for which two tones/delta-peaks can still be regarded as resolved, the minimum resolvable time-delay $\Delta\tau_{\min}$ and frequency difference $\Delta\nu_{\min}$, and the other resolution parameters based on them, for the WFT take the forms

$$\begin{aligned} &\Delta\nu_{\min} = \xi_2(\epsilon_r) - \xi_1(\epsilon_r), \quad \Delta\tau_{\min} \cong \tau_2(\epsilon_r) - \tau_1(\epsilon_r), \\ &\gamma_\omega \equiv \Delta\nu_{\min}^{-1}, \quad \gamma_t = \Delta\tau_{\min}^{-1}, \quad \gamma_{\omega t} \equiv \gamma_\omega\gamma_t = [\Delta\nu_{\min}\Delta\tau_{\min}]^{-1}, \end{aligned} \tag{2.32}$$

where $\tau_{1,2}(\epsilon)$ and $\xi_{1,2}(\epsilon)$ are defined in (2.26). Generally, accurate reconstruction might reasonably be assumed as being at 95 % precision, so one can set $\epsilon_r = 0.05$ in (2.32). The resolution characteristics of different windows are listed in Sect. 7.2.

The same approach straightforwardly extends to the WT case, where one applies similar considerations in terms of (2.27). Thus, it can be shown that two tones with frequency ratio $\frac{\nu+\Delta\nu}{\nu} = 1 + \Delta\nu/\nu = \frac{\xi_2(\epsilon)}{\xi_1(\epsilon)}$ are reconstructed from the WT with an overall relative error of around ϵ. This estimate is exact if $\hat{\psi}(\xi)$ is real, positive and symmetric on a logarithmic scale (such as the lognormal wavelet (2.19)), in which case the tones will always be separated at $\omega_x(t) = \exp[\log\nu + \log(\nu + \Delta\nu)] = \sqrt{\nu(\nu + \Delta\nu)}$. For the resolution of two delta-peaks, it follows from (2.21) that one should consider the ϵ-supports corresponding to $\psi^*(\omega t/\omega_\psi)$, so that the related error will be different at each frequency ω, characterized by ϵ calculated from $\omega\Delta\tau/\omega_\psi = \tau_2(\epsilon) - \tau_1(\epsilon)$. Hence, the resolution parameters (2.22), (2.23) for the WT are

$$\Delta\nu_{\min}(\nu) = \nu\Big(\frac{\xi_2(\epsilon_r)}{\xi_1(\epsilon_r)} - 1\Big), \quad \Delta\tau_{\min}(\nu) \cong \frac{\omega_\psi}{\nu}\big(\tau_2(\epsilon_r) - \tau_1(\epsilon_r)\big),$$
$$\gamma_\omega(\nu) \equiv [\Delta\nu_{\min}(\nu)]^{-1}, \quad \gamma_t(\nu) = [\Delta\tau_{\min}(\nu)]^{-1},$$
$$\gamma_{\omega t} \cong \Big[\omega_\psi\big(\tau_2(\epsilon_r) - \tau_1(\epsilon_r)\big)\log\frac{\xi_2(\epsilon_r)}{\xi_1(\epsilon_r)}\Big]^{-1}, \tag{2.33}$$

where $\tau_{1,2}(\epsilon)$ and $\xi_{1,2}(\epsilon)$ are defined in (2.27), and ϵ_r denotes the maximum allowable reconstruction error, which can be set to $\epsilon_r = 0.05$, similarly to that in (2.32). The resolution characteristics of different wavelets are listed in Sect. 7.2.

Summarizing, in contrast to the classic resolution measures (2.24), (2.25), the quantities in (2.32), (2.33) are very universal and have clear physical meaning, being related directly to the accuracy with which two time or frequency events can be recovered from the resultant TFR. In the context of the present work, where the main topic is the TFR-based decomposition of the signal (see Chaps. 3 and 4), such an approach seems to be the most relevant.

2.4 Practical Issues

In theory, one has infinite time and frequency scales, and both of these variables are continuous. In practice, however, everything is finite and discrete, which has specific consequences in terms of the resultant TFRs. In this section, the issues that arise while dealing with real signals are reviewed and studied.

2.4.1 Signal Preprocessing

To obtain a reliable TFR, an initial preprocessing of the signal should be performed. It consists of eliminating trends, followed by bandpass filtering in the frequency band of interest (for which the TFR is to be calculated). These two steps are considered below, with their effects being illustrated in Fig. 2.3.

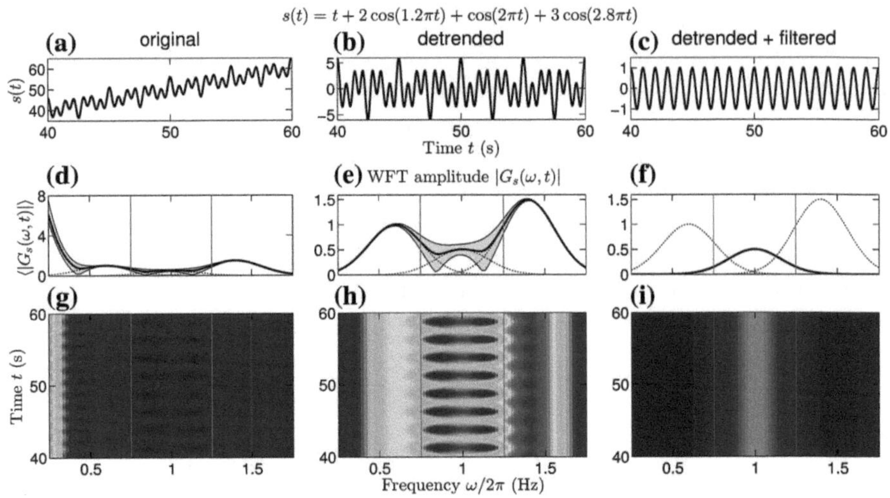

Fig. 2.3 Preprocessing and its effects on the resultant WFT (the case of the WT is qualitatively similar). **a** The original signal, consisting of the trend and three tones (shown at the *top*). **b** The same signal after detrending. **c** The same signal after detrending plus filtering in the band [0.75, 1.25] Hz. **d**, **e**, **f** The corresponding mean WFT amplitudes, with their 95 % ranges being indicated by *yellow regions*; *dotted lines* show the mean amplitudes for the WFTs of each tone (out of 3) separately. **g**, **h**, **i** The corresponding WFT amplitudes in the time-frequency plane, with the respective *color ranges* being the same as y-axis limits in (**d**, **e**, **f**). It is assumed that one is interested in calculating the TFR for the frequency range [0.75, 1.25] Hz, which is indicated by gray vertical lines in panels (**d**–**i**), but for completeness the resultant transforms are shown within wider ranges. The signal was sampled at 100 Hz for 100 s

2.4.1.1 Removing the Trends

If the signal contains a trend-like term (i.e. a term of the form Kt or more generally Kt^{α}), it can seriously corrupt the resultant TFR and complicate its interpretation. This is because trends make non-negligible contributions to the signal's spectral power in a wide frequency band. Thus, the FT of a trend existing for time T will be $\int_0^T Kte^{-i\xi t}dt = \frac{KT}{\xi}e^{-i\xi T} + \frac{K}{\xi^2}(e^{-i\xi T} - 1)$, which is proportional to the overall time of the trend and decays slowly with ξ. In general, any order contributions Kt^n, $n \in \mathbb{N}$ will have FT $\sim iT^n e^{-i\xi T}/\xi$ in the first order over ξ^{-1}. Therefore, it is clear from the frequency domain form of (2.8) and (2.13) that trends might seriously affect the representation of the other components in the TFR, as illustrated in Fig. 2.3d, g.

To avoid this, one should remove trends before doing any time-frequency analysis. To do so, one can subtract a simple linear fit of the data, which will eliminate the $\sim\xi^{-1}$ spectral contribution of any term $\sim t^n$, $n \geq 1$, changing it to $\sim\xi^{-2}$ for $n > 1$. It might be better, however, to subtract a third order polynomial fit, which will in addition fully remove the trends $\sim t^2, t^3$ and reduce the spectral power of the higher order terms to $\sim\xi^{-4}$; it will also eliminate to a large extent the step-increases in mean value that are sometimes present in real data. This approach is very simple and introduces minimal undesirable distortions to the signal's spectrum, as can be seen by comparing (d, g) and (e, h) in Fig. 2.3. Third order is suggested because such a polynomial has at most 3 zero crossings and thus cannot model more than one oscillation during the whole time-series; and since oscillations having less than one cycle over the signal time length cannot in principle be reliably studied in the TFR (see Sect. 2.4.4 below), one does not lose anything by filtering them out.

Remark 2.4.1 Sometimes detrending is performed by subtracting a moving average from the signal, which additionally filters out the low-frequency spectral content. However, due both to the properties of such a filter and its associated boundary effects, this procedure usually introduces more unwanted spectral distortions than subtraction of a simple polynomial fit followed by bandpass filtering (see below). In general, however, there exist many different approaches for trend removal [11], with the preferred choice being application dependent.

2.4.1.2 Bandpass Filtering

Using time-frequency analysis for a given signal, one is usually interested only in a particular frequency range (e.g. containing a chosen AM/FM component). At the same time, due to the peak broadening resulting from the TFR's finite frequency resolution, the frequency content of the signal *outside* the given range can significantly influence the TFR *inside* that range. This effect can be especially prominent when the spread of the window/wavelet function in the frequency domain is large. Therefore, one should always filter the signal in the frequency range of interest $[\omega_{\min}, \omega_{\max}]$ before or during the application of the TFR, i.e. set $\hat{s}(\xi < \omega_{\min}) = \hat{s}(\xi > \omega_{\max}) = 0$ in (2.8) and (2.13) (or use some other filter to remove spectral content lying outside the considered frequency range).

This issue is illustrated in Fig. 2.3e, h, where the WFT of the tone in a frequency range of interest (indicated by gray vertical lines) is seriously corrupted by the other two tones that lie near in frequency. As seen from Fig. 2.3f, i, filtration of the signal only within the band considered solves this problem, allowing for an accurate representation of the corresponding component. Note that, performed alone, bandpass filtering does not fully remove the influence of a trend in the frequency range under consideration. That is why trends should be removed first, as described in the previous subsection.

Remark 2.4.2 It should be noted, that even if the instantaneous frequency of the AM/FM component lies within the considered frequency range $[\omega_{\min}, \omega_{\max}]$, some of the related tones responsible for its amplitude/frequency modulation (see Eq. 2.6 and its discussion) might lie outside this range. In this case they will be filtered out, which will spoil the representation of the corresponding component to some extent. However, the probability that the related spectral content lies outside the considered band is not higher than the probability that there exists some unrelated components which, if unfiltered, might affect the components of interest to the same or a greater extent. Additionally, real signals are usually corrupted by noise, which can be viewed as many undesirable tones. Therefore, despite the possible drawbacks, it is generally preferable to filter than not to do so. Note also that there are many other filters [39, 53] that can be used instead of a simple bandpass filtering.

2.4.2 Frequency Discretization

In theory one has a continuous frequency variable ω, for which TFRs are calculated. In practice, however, the frequency axis is discretized, being partitioned into bins centered at the chosen discrete values $\omega = \omega_k$, and the WFT/WT is calculated only for these frequencies. As discussed previously (see Sect. 2.2), the WFT and WT have linear and logarithmic frequency resolutions, so that the discretization should also be performed on linear and logarithmic scales, respectively. It is convenient to take $\omega_k = (k - k_0)\Delta\omega$ (frequency bins $[(k - k_0 - 1/2)\Delta\omega, (k - k_0 + 1/2)\Delta\omega]$) for the WFT and $\omega_k/2\pi = 2^{(k-k_0)/n_v}$ (frequency bins $2\pi[2^{(k-k_0-1/2)/n_v}, 2^{(k-k_0+1/2)/n_v}]$) for the WT. The question is then what effects the choices of $\Delta\omega$ and n_v have on the resultant TFRs, and how to select these parameters appropriately.

Theoretically, the frequency resolution of the TFR is determined only by the window/wavelet properties (see Sect. 2.3). However, due to frequency discretization there also appears numerical frequency resolution, determined by the widths of the frequency bins, i.e. the choice of $\Delta\omega$ for the WFT or the number-of-voices n_v for the WT. It imposes an upper bound on the effective frequency resolution of the transform, which is equal to the minimum among the theoretical and numerical resolutions. Thus, if $\Delta\omega$ or $1/n_v$ is chosen to be too large, a few peaks in the TFR amplitude might be merged into one frequency bin, leading to an inability to distinguish between them and lowering the effective frequency resolution, as illustrated in Fig. 2.4. On the other hand, if $\Delta\omega$ or $1/n_v$ is selected to be too small, it cannot improve the frequency resolution beyond the theoretical maximum, but it will increase the computational cost due to requiring the calculation of the TFR at more frequencies. Therefore, $\Delta\omega$ and n_v should be selected so as to retain the original, theoretical frequency resolution, predicted from the chosen window/wavelet parameters. This maximizes the effective resolution and occurs when the numerical frequency resolution is the same as or better than the theoretical one. Obviously, the optimal values of $\Delta\omega$ and n_v depend on the chosen form of window/wavelet and its properties (see Fig. 2.4). Thus, while for $f_0 = 1$ a frequency step of $\Delta\omega/2\pi = 0.08$ is sufficient to distinguish between two nearby but theoretically resolved components (Fig. 2.4b), for $f_0 = 4$ it is already insufficient (Fig. 2.4c).

To avoid reducing the original frequency resolution by discretization, for the WFT the width of a frequency bin $\Delta\omega$ should be smaller than the theoretical $\Delta\nu_{\min}$ (see Sect. 2.3); while for the WT one has the same criterion but on a logarithmic frequency scale. In other words, one should break the minimum resolvable frequency difference (WFT) or ratio (WT) into a large enough number of segments. However, one is now interested in preserving the TFR's appearance more generally, i.e. not

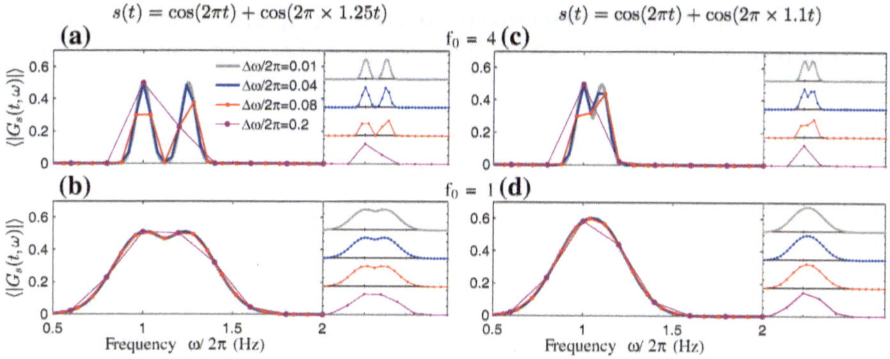

Fig. 2.4 Time-averaged WFT amplitudes for the signals $s(t) = \cos(2\pi t) + \cos(2.5\pi t)$ (**a**, **b**) and $s(t) = \cos(2\pi t) + \cos(2.2\pi t)$ (**c**, **d**), calculated using different resolution parameters f_0 (corresponding to rows) and frequency bin widths $\Delta\omega$; in each case, the *small insets* on the *right* show pictures for the considered $\Delta\omega$ individually. The situation for the WT is qualitatively the same

only for the reliably resolved components, but in relation to *any* components that can be distinguished in the TFR (even where they are substantially corrupted by interference). Thus, the optimal frequency bin widths should be smaller than $\Delta\nu_{\min}$ (2.32) and (2.33), determined for a relatively high error threshold ϵ_r; the latter is therefore set to an extreme value of $\epsilon_r = 0.5$. Based on these considerations, one can select an appropriate $\Delta\omega$ or n_v as

$$\begin{aligned}
&\textbf{WFT: } \Delta\omega = \frac{\xi_2(0.5) - \xi_1(0.5)}{N_b} \quad \big(\approx 1.35/f_0 N_b \text{ for Gaussian window (2.12)}\big),\\
&\textbf{WT: } n_v = \frac{N_b \log 2}{\log \xi_2(0.5) - \log \xi_1(0.5)} \quad \big(\approx 3.23 f_0 N_b \text{ for lognormal wavelet (2.19)}\big),
\end{aligned} \tag{2.34}$$

where $N_b > 1$ denotes the chosen number of bins into which the 50 %-support of the window/wavelet FT is divided (in the following $N_b = 10$ is used by default), and $\xi_{1,2}(\epsilon)$ are as defined in (2.26) and (2.27) for the WFT and WT, respectively.

The criterion (2.34) is based on (2.32) and (2.33), which are related to the resolution of two equal-amplitude tones. However, it is clear that, in order to be resolved, tones with different amplitudes (e.g. $s(t) = \cos\nu t + 0.1\cos(\nu + \Delta\nu)t)$ should have a bigger frequency difference than the tones of equal amplitudes. Therefore, $\Delta\omega$ and n_v are sufficient to resolve *any* two components if they are separated theoretically, as desired.

Remark 2.4.3 It should be noted, that even with the choice (2.34) it is still possible that at certain times close peaks will appear in the WFT/WT amplitude that are separated theoretically, but not practically, being merged due to insufficient numerical resolution. However, as mentioned previously, if the bin widths (2.34) are not small enough to resolve two tones, most of the time they will behave as a single component in the TFR, and it will be impossible to investigate them separately. Hence, attributing the corresponding peaks to one or few components will not make much difference.

2.4.3 Boundary Effects and Padding

Theoretically, one integrates over an infinite time or frequency axis while computing the WFT (2.8) and WT (2.13), but in practice the signal has a finite time duration and sampling frequency. Consequently, the resultant TFR becomes ill-defined near the signal's time borders (when t is close either to 0 or to the overall time-length T). Irrespectively of how this problem is tackled, it often leads to distortions of the TFR near both signal ends—*boundary effects*.

2.4.3.1 Padding and Its Schemes

Although this is not really the case in practice (see below), suppose for now that the WFT and WT are calculated using the convolution in the time domain in (2.8) and (2.13), respectively. The signal's positive part $s^+(\tau)$ is thus multiplied by $g(\tau - t)e^{-i\omega(\tau-t)}$ (WFT) or $\psi^*(\omega(\tau - t)/\omega_\psi)$ (WT) and integrated over τ. Therefore, one needs to devise a rule by which the integration can be performed when τ lies outside the signal's time limits, i.e. when $\tau < 0$ or $\tau > T$. Generally, this can be done by continuing the signal beyond its original time interval $[0, T]$ in some way, e.g. setting $s^+(\tau \notin [0, T]) = 0$ in (2.8), (2.13). In practice, the signal is padded at both ends using a particular convention, then the convolution is calculated with this padded signal, and after this only the part lying within the original time limits is taken. An appropriate padding can be constructed by many different schemes, and the most common/useful ones are listed below, with their effect on the resultant TFR being illustrated in Fig. 2.5. Note, that padding should be performed *after* the initial signal preprocessing, discussed in Sect. 2.4.1, has been completed.

Zero padding, when one pads the signal with zeros at both ends, represents the simplest and most predictable form of padding. This scheme effectively sets $s(\tau < 0) = s(\tau > T) = 0$ in (2.8), (2.13), "cutting" the convolution when τ goes beyond the signal's time limits. Its effect is illustrated in Fig. 2.5a, e: as can be seen, this type of padding does not introduce any new behavior (in contrast to a few other types, see below), always leading to a gradual decay of the TFR amplitude towards the time ends. This allows one to derive the expressions for boundary errors that arise in the case of zero padding, which will be done in the following.

Periodic padding is constructed by periodic continuation of the signal. It leads to quite unpredictable boundary effects, with the TFR near the time borders being affected by the signal's behavior at both ends. For example, as illustrated in Fig. 2.5b, the tone occurring at the beginning (end) has its "phantom" tail near the time end (beginning) in the WFT. Consequently, this type of padding does not represent the

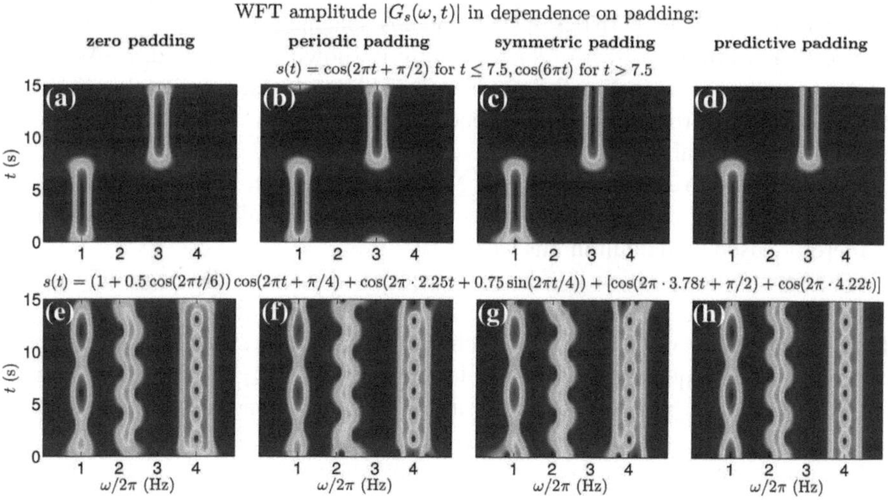

Fig. 2.5 Examples of the WFT amplitudes calculated using each of the four padding schemes discussed in the text (*columns*) for two different signals (*rows*): **a**, **e** zero padding; **b**, **f** periodic padding; **c**, **g** symmetric padding; **d**, **h** predictive padding. The effects of different padding strategies on the WT are qualitatively the same. The signals are indicated in the figure, and they were sampled at $f_s = 100$ Hz for $T = 15$ s

preferred choice, unless one knows that the signal is fully periodic over the whole interval, which almost never occurs in practice.

Symmetric padding whose effects are shown in Fig. 2.5c, g, is performed by reflecting the signal along $t = 0$ and $t = T$. It localizes boundary inaccuracies at each edge, thus solving the problem of the end-to-end influence that occurs for periodic padding. Indeed, comparing (c, g) and (b, f) of Fig. 2.5, it can be seen that the "phantom tails" near the boundaries have been removed. However, the effect of symmetric padding depends to a large extent on the phases of oscillations at $t = 0, T$, which are quite unpredictable. Thus, if the initial/end phase is not equal to zero or π, symmetric padding introduces a phase jump and leads to a splitting of the single peak in the TFR amplitude into several peaks near the time borders. As an example, in Fig. 2.5b the first oscillation has a phase shift of $\pi/2$ at $t = 0$ and thus is doubled, while the second one has zero shift at $t = T$ and thus is well represented.

Predictive padding, as implied by its name, continues the signal beyond its time limits by inferring/forecasting its past/future behavior. Thus, one tries to predict the values of the signal by assuming some model of the process generating the time-series and fitting data to this model to find its parameters. In the present case, one is mainly interested in continuing the signal in such a way as to best represent in the TFR its existing characteristics near the boundaries, and not in finding the unknown signal behavior beyond the available data, which is generally impossible. Because time-frequency analysis is devoted to studying the oscillatory properties of the data, it seems that the most relevant approach would be to forecast the signal based on its local spectral characteristics at both ends. A scheme for doing this is discussed in Sect. 7.4, while the effect of such padding is shown in Fig. 2.5d, h. As can be seen, for both signals in Fig. 2.5 predictive padding almost completely eliminates the boundary distortions. However, it does not represent an "ultimate cure", and for a complicated signal some errors might remain; they depend on the signal's structure and thus are hard to estimate in general. Nevertheless, in terms of its ability to reduce boundary effects, predictive padding usually outperforms all other schemes, and it is therefore used by default in this work.

The remaining question is how many values to pad. At the beginning of this section it was assumed for simplicity that the WFT/WT are calculated using the time domain forms of (2.8), (2.13). Such an approach, however, is computationally very expensive, requiring $O(N)$ convolutions for each $t_{n=1,\dots,N}$ and thus being of $O(N^2)$ in cost. In practice, therefore, the WFT and WT are calculated using the frequency domain convolutions in (2.8) and (2.13) which, utilizing the FFT algorithm, can be performed in $O(N \log N)$ computations (see Sect. 7.5). However, the discrete FT of the signal $\hat{s}(\xi_n)$, which is used in numerical convolution, represents the periodic spectrum estimate, being an exact FT of the periodically continued signal: $\hat{s}(\xi_n) = \int s(\text{mod}_T t)e^{-i\xi_n t}dt$. Thus, the absence of padding is in practice completely equivalent to periodic padding.

One therefore needs to add enough points at both signal ends to guarantee that, within the original signal time-length, the effects of implicit periodic continuation due to FFT-based convolution are small. This minimum number of padded values will obviously depend on the spread of the window/wavelet in time: the larger it is, the more values one should pad so that most of the window/wavelet is convoluted with padded values and not with the periodically-continued other end of the signal. Since the effective window/wavelet length in time (up to a predefined tolerance ϵ) can be expressed through its ϵ-supports $\tau_{1,2}(\epsilon)$ (2.26), (2.27), the minimum number of points $n_1^{(\text{min})}$ ($n_2^{(\text{min})}$) with which one should pad the signal for $t < 0$ ($t > T$) can be determined as

$$\begin{aligned} \textbf{WFT: } n_{1,2}^{(\text{min})}(\epsilon) &= f_s|\tau_{1,2}(\epsilon)|, \\ \textbf{WT: } n_{1,2}^{(\text{min})}(\epsilon) &= \frac{\omega_\psi}{\omega_{\text{min}}} f_s|\tau_{1,2}(\epsilon)|, \end{aligned} \tag{2.35}$$

where f_s is the sampling frequency of the signal, and ω_{min} denotes the minimum frequency for which the TFR is calculated. Thus, for the WT the wavelet is rescaled at each frequency, so the number of points needed to assure the specified precision ϵ also scales; to guarantee the accuracy for all frequencies, the maximum number of points is taken, which corresponds to the lowest frequency.

Additionally, the FFT algorithm requires the total length of the signal in samples to be a power of two. Therefore, in practice one should simultaneously assure both the power-of-two points and criterion (2.35), which can be done by padding the signal from originally N to N_p points, with n_1 values to the left ($t < 0$) and n_2 to the right ($t > T$), given by

$$N_p = \text{NextPowerOfTwo}[N + n_1^{(\min)}(\epsilon) + n_2^{(\min)}(\epsilon)] = N + n_1 + n_2,$$
$$n_{1,2} = \frac{n_{1,2}^{(\min)}(\epsilon)}{n_1^{(\min)}(\epsilon) + n_2^{(\min)}(\epsilon)}(N_p - N), \tag{2.36}$$

where $n_{1,2}^{(\min)}(\epsilon)$ are as defined in (2.35); by default the precision $\epsilon = 0.01$ is used.

2.4.3.2 Error Estimates and the Cone-of-influence

As discussed above, for periodic, symmetric and predictive padding the boundary effects very much depend on the signal's structure. For zero padding, however, they are quite universal and can thus be estimated. A simple and straightforward way to do so is to quantify the relative boundary inaccuracies $\epsilon_b(\omega, t)$ through the difference between the theoretical (2.10), (2.16) and "practical" WFT/WT of the single-tone signal $s(t) = \cos(\nu t + \varphi)$, calculated at the tone frequency ν. Then in the case of zero padding one obtains

WFT:

$$\begin{aligned}
\epsilon_b(\nu, t) \equiv & \frac{\left|G_s(\nu, t) - \widetilde{G}_s(\nu, t)\right|}{|G_s(\nu, t)|} \\
= & \frac{\left|\frac{1}{2}\int e^{i(\nu\tau+\varphi)} g(\tau - t) e^{-i\nu(\tau-t)} d\tau - \frac{1}{2}\int_0^T e^{i(\nu\tau+\varphi)} g(\tau - t) e^{-i\nu(\tau-t)} d\tau\right|}{\left|\frac{1}{2}\int e^{i(\nu\tau+\varphi)} g(\tau - t) e^{-i\nu(\tau-t)} d\tau\right|} \\
\leq & \frac{\left|\int_{-\infty}^0 g(\tau - t) d\tau\right| + \left|\int_T^\infty g(\tau - t) d\tau\right|}{\left|\int g(\tau) d\tau\right|} \\
= & |P_g(-t)| + |P_g(t - T)|,
\end{aligned} \tag{2.37}$$

WT:

$$\begin{aligned}
\epsilon_b(\nu, t) \equiv & \frac{\left|W_s(\nu, t) - \widetilde{W}_s(\nu, t)\right|}{|W_s(\nu, t)|} \\
= & \frac{\left|\frac{1}{2}\int e^{i(\nu\tau+\varphi)} \psi^*(\nu(\tau - t)/\omega_\psi) \frac{\nu d\tau}{\omega_\psi} - \frac{1}{2}\int_0^T e^{i(\nu\tau+\varphi)} \psi^*(\nu(\tau - t)/\omega_\psi) \frac{\nu d\tau}{\omega_\psi}\right|}{\left|\frac{1}{2}\int e^{i(\nu\tau+\varphi)} \psi^*(\nu(\tau - t)/\omega_\psi) \frac{\nu d\tau}{\omega_\psi}\right|} \\
\leq & \frac{\left|\int_{-\infty}^0 \psi^*(u - \nu t/\omega_\psi) e^{i\omega_\psi u} du\right| + \left|\int_T^\infty \psi^*(u - \nu t/\omega_\psi) e^{i\omega_\psi u} du\right|}{\left|\int \psi^*(u - \nu t/\omega_\psi) e^{i\omega_\psi u} du\right|} \\
= & |P_\psi(-\nu t/\omega_\psi)| + |P_\psi(\nu(t - T)/\omega_\psi)|,
\end{aligned} \tag{2.38}$$

where $\widetilde{G}_s(\omega, t)$ and $\widetilde{W}_s(\omega, t)$ denote respectively the WFT and WT obtained in practice, while $P_{g,\psi}(\tau)$ are as defined in (2.26), (2.27). Note that $\epsilon_b(\omega, t)$ (2.37),

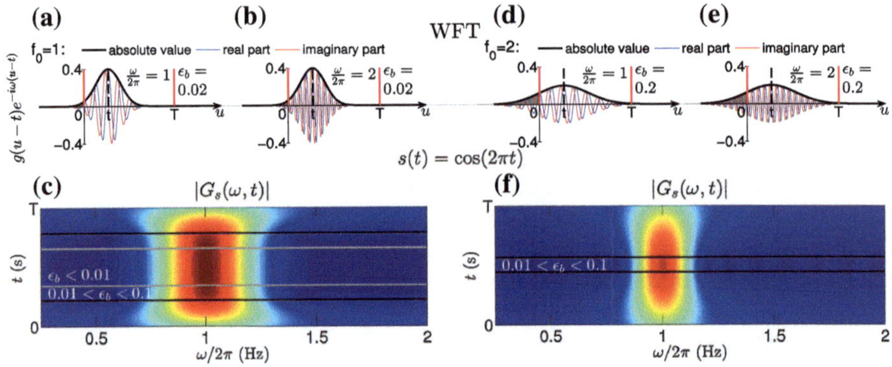

Fig. 2.6 **a, b, d, e** Examples of the time domain Gaussian window functions (2.12) (multiplied by the corresponding oscillatory part), with which the signal is convoluted while constructing the WFT (2.8), for different resolution parameters f_0 and frequencies ω; *thick vertical lines* indicate the time limits of the signal (*solid red*) and the time t at which the window is centered (*dashed black*); the area under the window modulus outside the time limits is *filled with gray* (note that, for a Gaussian window (2.12), $|g(t)| = g(t)$). **c, f** WFTs for the single tone signal $s(t) = \cos(2\pi t)$ in the same time limits $[0, T]$ as in (**a, b, d, e**), with *black* and *gray lines* enclosing the time-frequency regions where boundary errors (2.37) are small: $\epsilon_b < 0.1$ and $\epsilon_b < 0.01$, respectively. The signal was sampled at $f_s = 100$ Hz for $T = 7.5$ s, and in (**c, f**) padding with zeros was used

(2.38) does not depend on the phase-shift φ of the tone, as would be the case for the symmetric and periodic padding schemes.

It is clear that the boundary distortions in the TFR manifest themselves when a non-negligible portion of the window $g(\tau - t)e^{-i\omega(\tau-t)}$ or wavelet $\psi^*(\omega(\tau - t)/\omega_\psi)$, by which the signal $s^+(\tau)$ is multiplied and integrated in (2.8), (2.13), lies outside the time limits. This is true for any padding but, in the case of padding with zeros, the related errors take the simple forms (2.37), (2.38). From the latter it can be seen, that the dependence of the boundary inaccuracies on time and frequency is different for the WFT and WT, as illustrated in Figs. 2.6 and 2.7.

In the case of the WFT (Fig. 2.6), the boundary inaccuracies are independent of the frequency ω and depend only on time t: $\epsilon_b(\omega, t) = \epsilon_b(t)$. This is because ω controls the effective number of oscillations within the window, but not its spread in time, which is determined by the resolution parameter f_0. Hence, for each frequency the relative part of $g(\tau - t)$ lying outside the time limits (gray-shaded areas in Fig. 2.6a, b, d, e) is the same, so that for some (central) time range the WFT is well-behaved at all frequencies, as shown in Fig. 2.6c, f.

For the WT (Fig. 2.7), on the other hand, the wavelet is rescaled at each frequency ω, and so is the part of the wavelet outside the time limits (gray-shaded areas in Fig. 2.7a, b, d, e). As a result, the boundary errors depend on both frequency and time, and the region where the WT coefficients are determined with some predefined accuracy in terms of boundary effects takes the form of a cone, as shown in Fig. 2.7c, f.

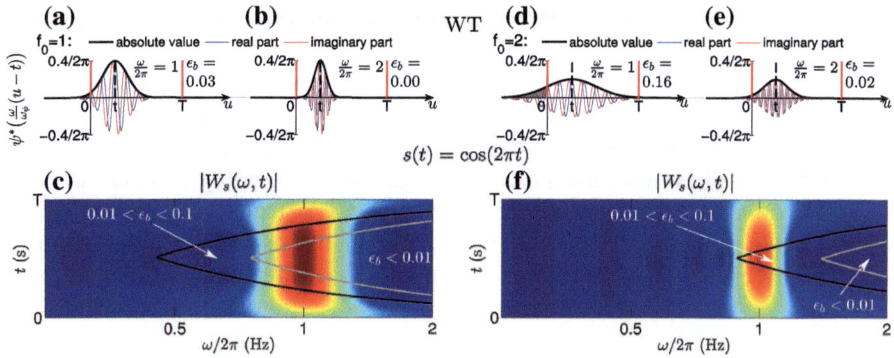

Fig. 2.7 Same as Fig. 2.6, but for the WT (2.13) with lognormal wavelet (2.19). **a**, **b**, **d**, **e** the wavelet function with which the signal is convoluted while constructing the WT (2.13); the area under the wavelet modulus outside the time limits is *filled with gray* (note that $|\psi(t)| \approx \psi(t)e^{-i\omega_\psi t}$ in the present case). **c**, **f** WTs for the single tone signal $s(t) = \cos(2\pi t)$ within the same time limits $[0, T]$ as in (**a**, **b**, **d**, **e**). The signal was sampled at $f_s = 100$ Hz for $T = 7.5$ s, and in (**c**, **f**) padding with zeros was used

The time-frequency region where the boundary errors are small in the current TFR is called the *cone-of-influence* (although this region has a truly conic form only for the WT):

$$\textbf{Cone-of-influence } \{\omega, t\}^{(\epsilon)}_{coi} : \ (\omega, t) \in \{\omega, t\}^{(\epsilon)}_{coi} \Leftrightarrow \epsilon_b(\omega, t) \le \epsilon, \tag{2.39}$$

where ϵ is the chosen accuracy threshold. Examples of the cones-of-influence for $\epsilon = 0.1$ and $\epsilon = 0.01$ have already been presented in Figs. 2.6c, f and 2.7c, f, where they are enclosed by black and gray lines, respectively. Based on the error estimates (2.37), (2.38), the cones-of-influence for the WFT and WT can be expressed through the corresponding window/wavelet ϵ-supports in time (2.26), (2.27) as

$$\begin{aligned}
\textbf{WFT: } & \{\omega, t\}^{(\epsilon)}_{coi} = \Big\{\big[\omega_{\min}, \omega_{\max}\big], \big[\delta t_1(\epsilon), T - \delta t_2(\epsilon)\big]\Big\}, \\
& \delta t_1(\epsilon) \le -\tau_1(\epsilon), \ \delta t_2(\epsilon) \le \tau_2(\epsilon), \\
\textbf{WT: } & \{\omega, t\}^{(\epsilon)}_{coi} = \Big\{\big[\omega_{\min}, \omega_{\max}\big], \big[\delta t_1(\epsilon, \omega), T - \delta t_2(\epsilon, \omega)\big]\Big\}, \\
& \delta t_1(\epsilon, \omega) \le -\omega_\psi \tau_1(\epsilon)/\omega, \ \delta t_2(\epsilon, \omega) \le \omega_\psi \tau_2(\epsilon)/\omega,
\end{aligned} \tag{2.40}$$

where $[\omega_{\min}, \omega_{\max}]$ is the frequency range in which the WFT/WT is calculated (the restrictions on this range and related issues are discussed in Sect. 2.4.4 below).

Because boundary effects can greatly influence the TFR behavior, especially for the WT, it is recommended that all characteristics (e.g. the WFT/WT mean amplitudes) be calculated using only TFR coefficients inside the cone-of-influence. The most appropriate padding scheme in this case is zero padding, as the corresponding boundary effects have a universal and well-defined form. In fact, the boundary errors

(2.37), (2.38) and the cone-of-influence (2.40) were rigorously estimated exclusively for this type of padding.

However, if one wants to extract some AM/FM component from the signal's TFR (the ways of doing this will be discussed in Chap. 3), then it should obviously be extracted for all time, and the consideration cannot be restricted to the cone-of-influence only. Thus, if all the TFR coefficients are to be used, then zero padding does not represent a good choice, because the WFT/WT near the boundaries will *surely* contain considerable errors. In this case predictive padding would be the most suitable, as it usually has the best performance in terms of reducing boundary effects.

2.4.4 TFR Frequency Range

For completeness, the restrictions on the frequency range $[\omega_{\min}, \omega_{\max}]$ over which to calculate the TFR, i.e. how high/low in frequency one can in principle go, should also be discussed. Consider a signal $s(t)$ sampled at f_s Hz for T seconds. Then one has

$$\omega_{\min}/2\pi \geq 1/T, \quad \omega_{\max}/2\pi \leq f_s/2. \tag{2.41}$$

The restriction on $\omega_{\max}$ follows from the Nyquist theorem, which states that oscillations with frequencies higher than half of the sampling frequency cannot be represented in a discrete time signal. The constraints on $\omega_{\min}$, on the other hand, are based on simple logic: it is clear that for a particular oscillation to be reliably studied (generally by any method), there should be at least one of its cycles within the signal.

However, from the statistical viewpoint, to reach any meaningful conclusions about the characteristics of the oscillatory process, such as its typical amplitude (as calculated e.g. from the time-averaged TFR amplitude), there should be at least 5-6 corresponding cycles within the signal [31], so that

$$\omega_{\min}^{(\text{stat})}/2\pi = 5/T. \tag{2.42}$$

Thus, although one can in principle estimate the properties of the components at lower frequencies, the resultant estimates might be highly untypical. Indeed, the number of cycles can be associated with a number of trials. For example, when testing the effects of some drug, one cannot base conclusions on only one subject (since the probability that the subject tested is an outlier is quite high); but if the same effects appear in 5-6 subjects, this suggests that they are quite common. Note also that, for statistical comparison of oscillatory properties between different data (e.g. as is done in [25, 45]), all these properties should be statistically meaningful in themselves, so that the related oscillations have frequencies $\geq \omega_{\min}^{(\text{stat})}$.

Finally, since the cone-of-influence for the WT contracts towards the lower frequencies (see Sect. 2.4.3), there exists also the minimal frequency $\omega_{\min}^{(\epsilon)}$ from the viewpoint of precision (while there is no such restriction in the case of the WFT).

The latter can be defined as the minimum frequency for which at least one WT coefficient is determined with accuracy ϵ in terms of the boundary effects. Using the estimates (2.40), one obtains

$$\textbf{WFT: } \omega_{\min}^{(\epsilon)} = -\infty; \qquad \textbf{WT: } \omega_{\min}^{(\epsilon)} \leq \omega_\psi(\tau_2(\epsilon) - \tau_1(\epsilon))/T. \tag{2.43}$$

Summarizing, it is quite safe to calculate and analyse the TFR in the range $[\omega_{\min}, \omega_{\max}]$ with $\omega_{\min} > \max[2\pi/T, \omega_{\min}^{(\epsilon)}]$ and $\omega_{\max} < 2\pi f_s/2$. However, if one needs to draw conclusions about the typical oscillatory parameters of the data, they should be based on the TFR within the cone-of-influence (see Sect. 2.4.3) and on at least 5 cycles of the related oscillations, so that $\omega_{\min} \geq \max[2\pi \times 5/T, \omega_{\min}^{(\epsilon)}]$.

Remark 2.4.4 Even if one is interested in the frequency range $[\omega_{\min}, \omega_{\max}]$ (so that the signal is filtered in this range on the stage of preprocessing, see Sect. 2.4.1), it might be useful to calculate the TFR for a larger interval (up to $(-\infty, \infty)$ for the WFT and $(0, \infty)$ for the WT). This allows to trace the "tails" of the TFR amplitude peaks located in the original range, which might be needed e.g. for a reliable parameter reconstruction by the direct method (see Sect. 3.2.2). For example, in Fig. 2.3f, i the considered frequency band—between the gray vertical lines—contains only the main part of the corresponding peak, with its tails occupying a wider range. If one wants to encompass the supports of all possible peaks contained in $[\omega_{\min}, \omega_{\max}]$ then, while bandpass filtering the signal in the original frequency band, its WFT/WT should be calculated within a slightly larger region $[\widetilde{\omega}_{\min}, \widetilde{\omega}_{\max}]$. To achieve relative accuracy ϵ, so that the new range contains the $(1 - \epsilon)$ part of the total area under any peak in the range of interest, one should use

$$\begin{aligned} \textbf{WFT: } [\widetilde{\omega}_{\min}, \widetilde{\omega}_{\max}] &= [\omega_{\min} + \xi_1(\epsilon), \omega_{\max} + \xi_2(\epsilon)], \\ \textbf{WT: } [\widetilde{\omega}_{\min}, \widetilde{\omega}_{\max}] &= \left[\omega_{\min}\frac{\xi_1(\epsilon)}{\xi_2(\epsilon)}, \omega_{\max}\frac{\xi_2(\epsilon)}{\xi_1(\epsilon)}\right], \end{aligned} \tag{2.44}$$

where the ϵ-supports $\xi_{1,2}(\epsilon)$ for the window and wavelet are defined in (2.26) and (2.27), respectively. Usually, however, this complication is not needed, unless the instantaneous frequency of some component of interest lies near $\omega_{\min}$ or $\omega_{\max}$.

2.5 Summary

This Chapter reviewed the main properties of the WFT and WT and discussed various issues related to their practical use, setting the foundations for the more advanced techniques discussed in the following Chapters. To summarize, the WFT and WT both represent particular projections of the signal onto the time-frequency plane, carried out as discussed in Sect. 2.2. They are mainly devoted to analysing signals of the form (2.5), i.e. those composed of a number of oscillatory components each of which satisfies the analytic approximation (2.4) (so that the corresponding amplitudes and phases are well-defined).

The WFT and WT, and generally any TFR, have finite time and frequency resolutions. That is, they can represent reliably only signals whose components do not vary too fast in time and are not too close to each other in frequency. By adjusting window/wavelet parameters (such as the resolution parameter f_0 in the case of Gaussian window and lognormal wavelet), one can vary resolution properties of the resultant transform. However, high time and high frequency resolutions are mutually exclusive, and increasing one of them will inevitably decrease the other.

The main difference between the WFT and WT lies in the type of resolution that they have: while the former resolves components based on absolute differences between their frequencies, the latter takes into account the corresponding ratios (or the difference between logarithms); in effect, time resolution of the WFT remains constant at all frequencies, while that of the WT increases with frequency. The resolution properties of both transforms, and the approaches that can be used for their quantification, were thoroughly considered in Sect. 2.3. In particular, it was shown that classical resolution measures appear to be highly non-universal and can give misleading results in certain cases (especially for the WT), and a more appropriate measures were introduced.

There are also many practical issues related to the application of the WFT and WT, which were discussed in detail in Sect. 2.4. Thus, before calculating the WFT or WT of the signal, it should first be detrended and filtered in the frequency range of interest (see Sect. 2.4.1). Next, one should also choose appropriate steps of frequency discretization, and the formula (2.34) was derived for this purpose. Finally, both TFRs suffer from the distortions near time boundaries; the expressions to estimate the related errors and the padding scheme which can be used to significantly reduce them were devised in Sect. 2.4.3. Section 7.5 provides complete algorithms for calculating the WFT and WT, while the commonest windows/wavelets and their properties are listed in Sect. 7.2.

References

1. P.S. Addison, *The Illustrated Wavelet Transform Handbook: Introductory Theory and Applications in Science, Engineering, Medicine and Finance* (IOP Publishing, Bristol, 2002)
2. L. Aguiar-Conraria, P.C. Magalhães, M.J. Soares, Cycles in politics: wavelet analysis of political time series. Am. J. Pol. Sci. **56**(2), 500–518 (2012)
3. A. Bandrivskyy, A. Bernjak, P.V.E. McClintock, A. Stefanovska, Wavelet phase coherence analysis: application to skin temperature and blood flow. Cardiovasc. Eng. **4**(1), 89–93 (2004)
4. E. Bedrosian, A product theorem for Hilbert transforms. Proc. IEEE **51**(5), 868–869 (1963)
5. D.S. Bloomfield, R.T.J. McAteer, B.W. Lites, P.G. Judge, M. Mathioudakis, F.P. Keenan, Wavelet phase coherence analysis: application to a quiet-sun magnetic element. Astrophys. J. **617**(1), 623 (2004)
6. B. Boashash, Estimating and interpreting the instantaneous frequency of a signal. I. Fundamentals. Proc. IEEE **80**(4), 520–538 (1992)

7. B. Boashash, Estimating and interpreting the instantaneous frequency of a signal. II. Algorithms and applications. Proc. IEEE **80**(4), 540–568 (1992)
8. B. Boashash, *Time Frequency Signal Analysis and Processing* (Elsevier, New York, 2003)
9. R.A. Carmona, W.L. Hwang, B. Torresani, Characterization of signals by the ridges of their wavelet transforms. IEEE Trans. Signal Proc. **45**(10), 2586–2590 (1997)
10. R.A. Carmona, W.L. Hwang, B. Torresani, Multiridge detection and time-frequency reconstruction. IEEE Trans. Signal Proc. **47**(2), 480–492 (1999)
11. R. Chandler, M. Scott, *Statistical Methods for Trend Detection and Analysis in the Environmental Sciences* (Wiley, New York, 2011)
12. H.-I. Choi, W.J. Williams, Improved time-frequency representation of multicomponent signals using exponential kernels. IEEE Trans. Acoust. Speech Signal Process. **37**(6), 862–871 (1989)
13. L. Cohen, Time-frequency distributions: a review. Proc. IEEE **77**(7), 941–981 (1989)
14. L. Cohen, *Time-Frequency Analysis*, vol. 778 (Prentice Hall PTR, New Jersey, 1995)
15. I. Daubechies, The wavelet transform, time-frequency localization and signal analysis. IEEE Trans. Inf. Theory **36**(5), 961–1005 (1990)
16. I. Daubechies, *Ten Lectures on Wavelets*, volume 61 of CBMS-NSF Regional Conference Series in Application Mathematics, SIAM, 1992
17. I. Daubechies, J. Lu, H. Wu, Synchrosqueezed wavelet transforms: an empirical mode decomposition-like tool. Appl. Comput. Harmon. Anal. **30**(2), 243–261 (2011)
18. N. Delprat, B. Escudie, P. Guillemain, R. Kronland-Martinet, P. Tchamitchian, B. Torrésani, Asymptotic wavelet and Gabor analysis: extraction of instantaneous frequencies. IEEE Trans. Inf. Theory **38**(2), 644–664 (1992)
19. D. Gabor, Theory of communication. J. IEEE **93**, 429–457 (1946)
20. A. Grinsted, J.C. Moore, S. Jevrejeva, Application of the cross wavelet transform and wavelet coherence to geophysical time series. Nonlinear Proc. Geophys. **11**(5/6), 561–566 (2004)
21. K. Gröchenig, *Foundations of Time-Frequency Analysis* (Springer, Berlin, 2001)
22. F. Hlawatsch, F. Auger, *Time-Frequency Analysis*, vol. 36 (Wiley, New York, 2010)
23. F. Hlawatsch, G.F. Boudreaux-Bartels, Linear and quadratic time-frequency signal representations. IEEE Signal Proc. Mag. **9**(2), 21–67 (1992)
24. N.E. Huang, Z. Wu, S.R. Long, K.C. Arnold, X. Chen, K. Blank, On instantaneous frequency. Adv. Adapt. Data Anal. **1**(02), 177–229 (2009)
25. D. Iatsenko, A. Bernjak, T. Stankovski, Y. Shiogai, P.J. Owen-Lynch, P.B.M. Clarkson, P.V.E. McClintock, A. Stefanovska, Evolution of cardio-respiratory interactions with age. Phil. Trans. R. Soc. Lond. A **371**(1997), 20110622 (2013)
26. D. Iatsenko, P.V.E. McClintock, A. Stefanovska, On the extraction of instantaneous frequencies from ridges in time-frequency representations of signals [preprint—arXiv:1310.7276], 2014
27. D. Iatsenko, P.V.E. McClintock, A. Stefanovska, Linear and synchrosqueezed time-frequency representations revisited: Overview, standards of use, reconstruction, resolution, concentration, and algorithms. Dig. Signal Proc. (in press), 2015, doi:10.1016/j.dsp.2015.03.004
28. J. Jamšek, M. Paluš, A. Stefanovska, Detecting couplings between interacting oscillators with time-varying basic frequencies: Instantaneous wavelet bispectrum and information theoretic approach. Phys. Rev. E **81**(3), 036207 (2010)
29. J. Jamšek, A. Stefanovska, P.V.E. McClintock, Wavelet bispectral analysis for the study of interactions among oscillators whose basic frequencies are significantly time variable. Phys. Rev. E **76**(4), 046221 (2007)
30. G. Kaiser, *A Friendly Guide to Wavelets* (Birkhäuser, Boston, 1994)
31. L. Keselbrener, S. Akselrod, Selective discrete Fourier transform algorithm for time-frequency analysis: method and application on simulated and cardiovascular signals. IEEE Trans. Biomed. Eng. **43**(8), 789–802 (1996)
32. P. Kumar, E. Foufoula-Georgiou, Wavelet analysis for geophysical applications. Rev. Geophys. **35**(4), 385–412 (1997)
33. D. Labat, Recent advances in wavelet analyses: part 1. A review of concepts. J. Hydrol. **314**(1), 275–288 (2005)

34. D. Labat, J. Ronchail, J.L. Guyot, Recent advances in wavelet analyses: part 2. Amazon, Parana, Orinoco and Congo discharges time scale variability. J. Hydrol. **314**(1), 289–311 (2005)
35. J.P. Lachaux, A. Lutz, D. Rudrauf, D. Cosmelli, M. Le van Quyen, J. Martinerie, F. Varela, Estimating the time-course of coherence between single-trial brain signals: an introduction to wavelet coherence. Clin. Neurophysiol. **32**(3), 157–174 (2002)
36. M. Malik, Heart rate variability. Ann. Noninvas. Electr. **1**(2), 151–181 (1996)
37. S. Mallat, *A Wavelet Tour of Signal Processing*, 3rd edn. (Academic Press, Burlington, 2008)
38. J. Morlet, Sampling theory and wave propagation, in *Issues on Acoustic Signal/Image Processing and Recognition*, vol. I, NATO ASI series, ed. by C.H. Chen (Springer, Berlin, 1983), pp. 233–261
39. A.V. Oppenheim, R.W. Schafer, J.R. Buck, *Discrete-time signal processing*, vol. 2. (Prentice Hall, New Jersey, 1989)
40. B. Picinbono, On instantaneous amplitude and phase of signals. IEEE Trans. Signal Proc. **45**(3), 552–560 (1997)
41. S. Qian, D. Chen, Joint time-frequency analysis. IEEE Signal Proc. Mag. **16**(2), 52–67 (1999)
42. A.W. Rihaczek, Signal energy distribution in time and frequency. IEEE Trans. Inf. Theory **14**(3), 369–374 (1968)
43. E. Sejdić, I. Djurović, J. Jiang, Time-frequency feature representation using energy concentration: an overview of recent advances. Dig. Signal Proc. **19**(1), 153–183 (2009)
44. L.W. Sheppard, A. Stefanovska, P.V.E. McClintock, Testing for time-localised coherence in bivariate data. Phys. Rev. E **85**, 046205 (2012)
45. Y. Shiogai, A. Stefanovska, P.V.E. McClintock, Nonlinear dynamics of cardiovascular ageing. Phys. Rep. **488**, 51–110 (2010)
46. L. Stankovic, M. Dakovic, T. Thayaparan, *Time-Frequency Signal Analysis with Applications* (Artech House, Norwood, 2013)
47. A. Stefanovska, Coupled oscillators: complex but not complicated cardiovascular and brain interactions. IEEE Eng. Med. Biol. Mag. **26**(6), 25–29 (2007)
48. A. Stefanovska, M. Bračič, Physics of the human cardiovascular system. Contemp. Phys. **40**(1), 31–55 (1999)
49. A. Stefanovska, M. Bračič, H.D. Kvernmo, Wavelet analysis of oscillations in the peripheral blood circulation measured by laser Doppler technique. IEEE Trans. Bio. Med. Eng. **46**(10), 1230–1239 (1999)
50. G. Thakur, E. Brevdo, N.S. Fuckar, H.-T. Wu, The Synchrosqueezing algorithm for time-varying spectral analysis: robustness properties and new paleoclimate applications. Sig. Process. **93**(5), 1079–1094 (2013)
51. C. Torrence, G.P. Compo, A practical guide to wavelet analysis. Bull. Am. Meteorol. Soc. **79**(1), 61–78 (1998)
52. D. Vakman, On the analytic signal, the Teager-Kaiser energy algorithm, and other methods for defining amplitude and frequency. IEEE Trans. Sig. Proc. **44**(4), 791–797 (1996)
53. M. Vetterli, J. Kovačević, *Wavelets and Subband Coding*, vol. 87 (Prentice Hall, Englewood Cliffs, 1995)
54. J.D. Ville et al., Théorie et Applications de la Notion de Signal Analytique. Câables et transmission **2**(1), 61–74 (1948)
55. M. Wacker, H. Witte, Time-frequency techniques in biomedical signal analysis. A tutorial review of similarities and differences. Method. Inform. Med. **52**(4), 279–296 (2013)
56. E. Wigner, On the quantum correction for thermodynamic equilibrium. Phys. Rev. **40**(5), 749 (1932)

Chapter 3
Extraction of Components from the TFR

The main purpose of the time-frequency analysis can be formulated as the identification and quantification of the AM/FM components present in the signal. Thus, with the help of TFRs one can distinguish the individual components present in the signal and estimate their time-dependent properties, such as the corresponding instantaneous amplitudes, phases and frequencies. In this respect TFRs can be used to decompose a signal into its constituent components, or just to recover some particular components of interest [1, 4–9, 14]. This can be done in two steps. First, the component of interest should be identified, i.e. in the current TFR the time-frequency region where it is concentrated should be found. Then, knowing this region, one can estimate the parameters of the corresponding component using a specific reconstruction method. The present Chapter discusses these two steps, their implementation and related issues.

3.1 Identification of the Components

If the construction of the TFR is well-matched to the signal's structure, then each AM/FM component will appear as a "curve" in the time-frequency plane (e.g. see Fig. 2.1), formed by a unique sequence of TFR amplitude peaks—ridge points. To identify the component, one therefore needs to extract the corresponding curve, i.e. find the appropriate peak sequence by which this component is represented in the WFT/WT.

3.1.1 Ridge Curve and Its Extraction

In practice, a typical signal has the form (2.5), consisting of many AM/FM components $x_i(t)$ and some noise. If the noise is not very strong, and the resolution properties of the WFT/WT (determined by the parameters of window/wavelet used) are appropriate for the characteristic time modulation and frequency separation between the

D. Iatsenko, *Nonlinear Mode Decomposition*,
Springer Theses, DOI 10.1007/978-3-319-20016-3_3

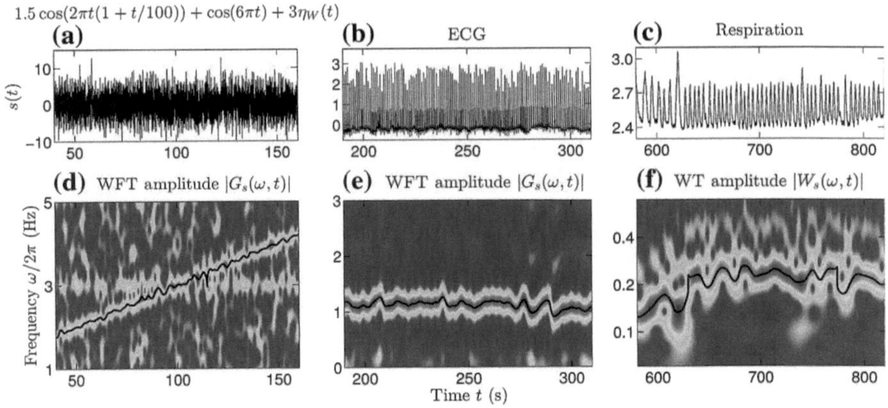

Fig. 3.1 Examples of the extracted ridge curves $\omega_p(t)$ for different signals: **a** simulated signal composed of one chirp, one tone and noise, as specified by equation above the panel ($\eta_W(t)$ denotes unit-deviation Gaussian *white* noise); **b** human electrocardiogram (ECG) signal (3-lead, with electrodes on shoulders and lowest left rib, see e.g. [12]); **c** human respiration signal (measured by belt, see e.g. [8]). **d, e, f** the TFRs of the signals shown in the *upper panels*; in each case, the *black line* shows the extracted *ridge curve* $\omega_p(t)$ (i.e. the sequence of TFR amplitude peaks) corresponding to the dominant component. The signals were sampled at 40 Hz for **a** 200s; **b, c** 1800s

components then, at each time, there is a unique peak in the TFR amplitude for each $x_i(t)$. The sequence of peaks associated with a particular component will be referred to as its *ridge curve*, while the corresponding frequency profile will be denoted as $\omega_p(t)$. It is usually represented by the peaks closest to the component's actual frequency, but the latter is almost never known a priori for real signals. Examples of the extracted $\omega_p(t)$ are presented in Fig. 3.1.

The problem of ridge curve extraction lies in selecting from among all possible combinations the sequence of peaks that corresponds to a single component. This is not a trivial issue, since in real cases there are often many peaks in the TFR amplitude at each time, and their number often varies. In such circumstances it can be unclear which peak corresponds to which component, and which are just noise-induced artifacts. Practically, it is convenient to extract the most dominant among $x_i(t)$ in (2.5), i.e. the one having the largest $\langle x_i^2(t)\rangle$. One can then reconstruct the corresponding component (see Sect. 3.2 below), subtract it from the signal and repeat the procedure to find the other components.

In what follows, the ridge points (i.e. positions of the amplitude peaks at each time), their numbers and the corresponding TFR amplitudes will be denoted as $\nu_m(t)$, $N_p(t)$ and $Q_m(t)$, respectively:

$$\begin{aligned}
&\nu_m(t):\ \begin{cases}\left[\partial_\omega |H_s(\omega,t)|\right]_{\omega=\nu_m(t)} = 0,\\ \left[\partial_\omega^2 |H_s(\omega,t)|\right]_{\omega=\nu_m(t)} < 0,\end{cases}\\
&Q_m(t) \equiv |H_s(\nu_m(t),t)|,\\
&m = 1,\ldots,N_p(t),
\end{aligned} \tag{3.1}$$

where $H_s(\omega, t)$ stands for the WFT $G_s(\omega, t)$ (2.8) or the WT $W_s(\omega, t)$ (2.13), depending on which TFR type is used. Clearly, the ridge curve can now be expressed as $\omega_p(t) = \nu_{m_c(t)}(t)$, where $m_c(t)$ is the sequence of the selected peak indices at each time t, which one needs to find. Note, that the number of peaks $N_p(t)$ can vary in time and in real cases is often greater than the number of components present in the signal, e.g. with the additional peaks being attributable to noise.

Remark 3.1.1 Since in practice the frequency scale for the WFT/WT is discretized, the positions of the peaks $\nu_m(t)$ also take discrete values. As a result, e.g. the time-derivative $d\omega_p(t)/dt$ cannot be reliably calculated, because its numerical estimate become "quantized" in steps determined both by the width of the frequency bins and the signal sampling frequency f_s. Thus, given a high enough f_s, the numerical $d\omega_p(t)/dt$ will be zero for most of the time (and exceedingly high at some moments). This greatly restricts the choice of techniques that can be used for curve extraction. Moreover, the performance of the corresponding methods might depend on numerical parameters (e.g. the sampling frequency) in this case. To avoid these adverse effects, the peak positions $\nu_m(t)$ (as well as their amplitudes $Q_m(t)$) are estimated more precisely using three-point parabolic interpolation for each peak (see (3.6) below). Both $\nu_m(t)$ and the numerical $d\omega_p(t)/dt$ then become continuous (rather than discretized), making all procedures more meaningful and universal.

The procedures that can be used for ridge curve extraction have been discussed, developed and compared in [9] (see also [1, 4, 5, 8]). Generally, $\omega_p(t)$ can be found as the trajectory maximizing some path functional of the ridge amplitudes and frequencies:

$$\begin{aligned}
&\{\omega_p(t_1), \dots, \omega_p(t_N)\} = \{\nu_{m_c(t_1)}(t_1), \dots, \nu_{m_c(t_N)}(t_N)\}, \ \{m_c(t_1), \dots, m_c(t_N)\} \\
&= \operatorname*{argmax}_{\{m_1, m_2, \dots, m_N\}} \sum_{n=1}^{N} F\big[t_n, \{Q_{m_1}(t_1), \dots, Q_{m_N}(t_N)\}, \{\nu_{m_1}(t_1), \dots, \nu_{m_N}(t_N)\}\big],
\end{aligned} \tag{3.2}$$

where the optimization is performed over all possible sequences of peak numbers $\{m_1, m_2, \dots, m_N\}$, and $F[\dots]$ denotes the chosen functional of the current discrete time t_n, the whole ridge frequency profile $\{\nu_{m(t)}(t)\}$ and the corresponding amplitude profile $\{Q_{m(t)}(t)\}$. Although such an approach appears to be computationally very expensive, it turns out that, if the functional at each time depends on only a finite number of the consecutive points rather than the full history, then using a dynamic programming algorithm [2, 3] one can select the optimal path in terms of (3.2) in $O(N)$ computations [9]. The algorithm for doing this is described in Sect. 7.6.

Remark 3.1.2 Note that, instead of the path optimization method (3.2), one can use a simple one-step optimization, selecting $m_c(t_n)$ as that maximizing $F[t_n, \dots]$ calculated using amplitude and frequency profiles formed from the points selected at previous steps. However, as investigated in [9], this approach gives results much worse than those obtained with (3.2) because, in the latter case, all the trajectories are effectively explored and therefore a more appropriate curve can be found.

3.1.2 Main Procedure

The principal question is what $F[\ldots]$ to use in (3.2). Obviously, it should favor the higher amplitudes $Q_m(t_n)$. Traditionally, it is also complemented with a penalization term, which suppresses the frequency differences between the subsequent ridges. However, such an approach will restrain any frequency variability, irrespectively of whether this variability is characteristic of some component or not. Therefore, it will be highly non-universal, working well for one type of signal and failing for the other (given the same method parameters). Thus, e.g. for chirps $x(t) \sim \cos(\nu t + at^2)$ it is clear that one should penalize not the frequency jumps, but their difference from the actual frequency growth rate.

To make the scheme adaptive, the parameters of the functional should be matched to the properties of the component being extracted, such as typical variations of its instantaneous frequency. The latter can be characterized by the averages and standard deviations of the ridge frequencies and their differences. Based on these characteristics, one can construct an adaptive functional by suppressing not the absolute frequency jumps, but the relative deviations of the component's frequency and its derivative from their typical values. It can be chosen as

$$
\begin{aligned}
&\textbf{WFT:}\ \big(\widetilde{\omega}_p(t_n) \equiv \nu_{m_n}(t_n)\big)\\
&F[t_n, \widetilde{\omega}_p(t)] \equiv \log|G_s(\widetilde{\omega}_p(t_n), t_n)| - p_1\left|\frac{\Delta\widetilde{\omega}_p(t_n) - \langle\Delta\widetilde{\omega}_p(t)\rangle}{\mathrm{std}[\Delta\widetilde{\omega}_p(t)]}\right|^{n_1}\\
&\qquad - p_2\left|\frac{\widetilde{\omega}_p(t_n) - \langle\widetilde{\omega}_p(t)\rangle}{\mathrm{std}[\widetilde{\omega}_p(t)]}\right|^{n_2},\\
&\textbf{WT:}\ \big(\widetilde{\omega}_p(t_n) \equiv \nu_{m_n}(t_n)\big)\\
&F[t_n, \widetilde{\omega}_p(t)] \equiv \log|W_s(\widetilde{\omega}_p(t_n), t_n)| - p_1\left|\frac{\Delta\log\widetilde{\omega}_p(t_n) - \langle\Delta\log\widetilde{\omega}_p(t)\rangle}{\mathrm{std}[\Delta\log\widetilde{\omega}_p(t)]}\right|^{n_1}\\
&\qquad - p_2\left|\frac{\log\widetilde{\omega}_p(t_n) - \langle\log\widetilde{\omega}_p(t)\rangle}{\mathrm{std}[\log\widetilde{\omega}_p(t)]}\right|^{n_2},
\end{aligned}
\tag{3.3}
$$

where $\Delta\widetilde{\omega}_p(t_n) \equiv \widetilde{\omega}_p(t_n) - \widetilde{\omega}_p(t_{n-1})$ and $\Delta\log\widetilde{\omega}_p(t_n) \equiv \log\widetilde{\omega}_p(t_n) - \log\widetilde{\omega}_p(t_{n-1})$; for simplicity everything is expressed through $\widetilde{\omega}_p(t_n) \equiv \nu_{m_n}(t_n)$, and (3.2) takes the form $\omega_p(t) = \underset{\widetilde{\omega}_p(t)}{\operatorname{argmax}} \sum_{n=1}^{N} F[t_n, \widetilde{\omega}_p(t)]$. Note that, to be consistent with the resolution properties of the respective TFRs (see Sect. 2.3), the frequency variables are considered on a linear scale for the WFT and on a logarithmic scale for the WT. The parameters $n_{1,2}$ and $p_{1,2}$ in (3.3) determine the laws and the extents of suppression of the corresponding relative deviations; by default their values are set to $n_{1,2} = 1$, $p_{1,2} = 1$, because such a choice was found to be quite universal and seems to work well in the majority of cases [9].

Remark 3.1.3 In (3.3), one can choose another function of the ridge amplitudes instead of the logarithm, e.g. simply its squared value; however, the logarithm seems to be the most appropriate because, in this case, the path functional (3.2) depends on the product of all amplitudes and thus can be significantly influenced even by a single "wrong" point, making selection of the latter less probable.

By maximizing the path integral (3.2) based on the functional (3.3), one in fact tries to extract the curve which is most consistent with itself. Thus, the strength of the respective frequency variations becomes unimportant, and it is only their agreement and similarity at different times that matters. Furthermore, by suppressing the relative deviations of the component's frequency from its mean, the functional (3.3) stabilizes the curve in its characteristic frequency range, thus lowering the possibility that it will "escape" and switch to another component.

However, the functional (3.3) depends on the whole time-evolution of $\omega_p(t)$, so the path optimization (3.2) cannot be performed using a fast $O(N)$ algorithm described in Sect. 7.6. Nevertheless, one can approach the approximate optimum curve $\omega_p(t)$ iteratively. Namely, given some initial guess $\omega_p^{(0)}(t)$, one calculates the corresponding averages and standard deviations, fixes them in (3.3) and, using in (3.2) the resultant functional (for which the path optimization algorithm discussed in Appendix 7.6 is applicable), extracts the newer profile $\omega_p^{(1)}(t)$. Then the (fixed) averages and deviations are updated to those for the $\omega_p^{(1)}(t)$ and, based on them, the next approximation $\omega_p^{(2)}(t)$ is found. The procedure is repeated until the curves obtained in two consecutive iterations fully coincide, indicating convergence (which is usually achieved within only a few iterations).

The initial guess $\omega_p^{(0)}(t)$ can be chosen as a simple maximum-based curve, formed by the positions of the highest peaks at each time. However, because such a trajectory might be composed from the parts of curves belonging to different components, it is better to use not the averages and deviations, but the medians $\mathfrak{m}[\ldots]$ and 50% ranges $\mathfrak{s}[\ldots]$:

$$\langle\ldots\rangle \to \mathfrak{m}[\ldots] = \mathrm{perc}_{0.5}[\ldots], \quad \mathrm{std}[\ldots] \to \mathfrak{s}[\ldots] = \mathrm{perc}_{0.75}[\ldots] - \mathrm{perc}_{0.25}[\ldots], \tag{3.4}$$

where $\mathrm{perc}_p[f(t)]$ denotes the pth quantile of $f(t)$. The medians and ranges are generally more universal than usual means and standard deviations, so for all calculations we do the replacement (3.4) in (3.3). Note, however, that such an approach works well only in conjunction with peak interpolation (see Remark 3.1.1), being useful only if $\omega_p(t)$ take continuous values; otherwise, the usual averages/deviations should be used.

Apart from (3.3), various other functionals were studied in [9]. However, the choice (3.3) was found to greatly outperform all the other ones. Being highly adaptive, it seems to work well in the majority of cases, e.g. all $\omega_p(t)$ presented in Fig. 3.1 for various signals were found by this scheme. Thus, the curve extraction procedure based on (3.2) and (3.3) with the default parameters $n_{1,2} = 1$, $p_{1,2} = 1$ appears to be of almost universal utility, being a type of "just apply" method which does not require any tuning from the user.

Remark 3.1.4 In addition to frequency and its difference, one can also suppress the relative deviations of the component's other parameters (the amplitude and its time-derivative, higher order frequency differences etc.), which can be done by introducing the corresponding terms into (3.3). However, the performance of the original functional is already very good [9], so there is no need to complicate it.

3.2 Estimation of the Component's Parameters

Having extracted the ridge curve $\omega_p(t)$, the question which immediately arises is how to find the instantaneous amplitude $A(t)$, phase $\phi(t)$ and frequency $\nu(t)$ of the associated component. These parameters can be reconstructed by two methods: ridge and direct. Both these approaches, as well as the differences between the corresponding estimates, are considered below. Note, that all TFR methods by definition aim to reconstruct the analytic approximation (2.4) to the component parameters. Therefore, in rare cases when the analytic estimates are inaccurate, all TFR-based estimates will be inaccurate to the same or higher extent.

3.2.1 Ridge Reconstruction

The component's parameters can be reconstructed directly from the TFR at the ridge points [7, 11, 13], which will be referred to as the *ridge reconstruction*. Thus, for the single tone signal $s(t) = A\cos(\nu t + \varphi)$ it follows from (2.10), (2.16) that the TFR amplitude at each time will be peaked at the tone frequency, and that the tone amplitude and phase can be perfectly reconstructed from the TFR value at any frequency as $Ae^{i(\nu t+\varphi)} = 2G_s(\omega, t)/\hat{g}(\omega - \nu)$ (WFT) or $Ae^{i(\nu t+\varphi)} = 2W_s(\omega, t)/\hat{\psi}^*(\omega_\psi \nu/\omega)$ (WT). Generalizing such an approach to the case of any AM/FM component, one obtains the ridge reconstruction formulas:

$$\begin{aligned}
&\textbf{ridge[WFT]:}\ \nu(t) = \omega_p(t),\ A(t)e^{i\phi(t)} = \frac{2G_s(\omega_p(t), t)}{\hat{g}(0)},\\
&\textbf{ridge[WT]:}\ \nu(t) = \omega_p(t),\ A(t)e^{i\phi(t)} = \frac{2W_s(\omega_p(t), t)}{\hat{\psi}^*(\omega_\psi)}.
\end{aligned} \tag{3.5}$$

Note that, for the WT, the expressions (3.5) should be modified if any other normalization than (2.13) is used (see Remark 2.2.2).

The expressions in (3.5) are derived for a continuous frequency scale, but in practice the TFR is calculated at discrete frequencies $\omega_k = (k - k_0)\Delta\omega$ (WFT) or $\omega_k/2\pi = 2^{(k-k_0)/n_v}$ (WT). If the corresponding estimates are discretized "directly", then they will contain errors proportional to $\Delta\omega$ (WFT) or n_v^{-1} (WT), which might be considerable. These errors can be significantly reduced by using the quadratic interpolation to better locate the position of the peak. Then, denoting as $k_p(t)$ the frequency bin corresponding to the ridge frequency $\omega_p(t)$, the discretized version of (3.5) takes the form

$$\textbf{ridge[WFT]:}\ \nu(t) = \omega_{k_p(t)}(t) + \delta\nu_d(t),\ A(t)e^{i\phi(t)} = \frac{2G_s(\omega_{k_p(t)}(t), t)}{\hat{g}(-\delta\nu_d(t))},$$

$$\delta\nu_d(t) = \frac{\Delta\omega}{2}\frac{a_3 - a_1}{2a_2 - a_1 - a_3},\ a_{\{1,2,3\}} \equiv |G_s(\omega_{\{k_p(t)-1, k_p(t), k_p(t)+1\}})|,$$

$$\textbf{ridge[WT]:}\ \nu(t) = \omega_{k_p(t)}(t)e^{\delta\log\nu_d(t)},\ A(t)e^{i\phi(t)} = \frac{2W_s(\omega_p(t), t)}{\hat{\psi}^*(\omega_\psi e^{\delta\log\nu_d(t)})},$$

$$\delta\log\nu_d(t) = \frac{n_v^{-1}\log 2}{2}\frac{a_3 - a_1}{2a_2 - a_1 - a_3},\ a_{\{1,2,3\}} \equiv |W_s(\omega_{\{k_p(t)-1, k_p(t), k_p(t)+1\}})|, \tag{3.6}$$

with the expressions for $A(t)e^{i\phi(t)}$ being obtained based on the WFT and WT of a single tone signal (see discussion preceding (3.5)). For such a signal it can be shown [10], that the discretization errors of the estimates (3.6) are $O(\Delta\omega^2)$ (WFT) or $O(n_v^{-2})$ (WT) for $\nu(t)$ and $\phi(t)$, and $O(\Delta\omega^3)$ (WFT) or $O(n_v^{-3})$ for $A(t)$. These errors are negligible if $\Delta\omega$ and n_v are selected based on the criteria (2.34). Note, however, that the above estimates assume that the first derivatives of the window and wavelet FTs are continuous at their peaks, i.e. $\hat{g}''(0)$ and $\hat{\psi}''(\omega_\psi)$ are finite; otherwise, all the errors become proportional to the first order of $\Delta\omega$ or n_v^{-1} [10].

3.2.2 *Direct Reconstruction*

Another approach is to reconstruct the component's parameters from the whole time-frequency region where it is concentrated in the current TFR. This region will be called the *time-frequency support* (TFS) of the component, and will be denoted as $[\omega_-(t), \omega_+(t)]$. Given the associated ridge curve $\omega_p(t)$, TFS can be defined as the widest region of unimodal non-zero amplitude around the corresponding peaks at each time:

Time-frequency support (TFS) $[\omega_-(t), \omega_+(t)]$**:**

$$\forall\omega \in [\omega_-(t), \omega_+(t)] :\ \text{sign}(\omega - \omega_p(t))\partial_\omega|H_s(\omega, t)| \leq 0,\ |H_s(\omega, t)| > 0, \tag{3.7}$$

where $|H_s(\omega, t)|$ denotes the chosen TFR (WFT $G_s(\omega, t)$ or WT $W_s(\omega, t)$), and sign$(\ldots)$ is the sign function.

Indeed, if two components interfere in the TFR, then one expects them to be nearly-optimally separated at the frequency where the amplitude minimum between the two corresponding peaks occurs, indicating that one component contribution was overcome by the other one. This can be shown rigorously (see Sect. 2.3) for the WFT and WT of the sum of two equal-amplitude tones (2.20) in the case of a real-positive and unimodal window and wavelet FTs satisfying $\hat{g}(-\xi) = \hat{g}(\xi), \forall\xi$ (e.g. Gaussian window) and $\hat{\psi}(a\omega_\psi) = \hat{\psi}(a^{-1}\omega_\psi), \forall a > 0$ (e.g. lognormal wavelet), respectively. In any case, the definition (3.7) is the most straightforward and intuitively clear one, and there are arguably no other reasonable choices. Examples of time-frequency supports are shown in Fig. 3.2 for the WFT (for the WT all is qualitatively the same).

Remark 3.2.1 It should be noted, that the definition (3.7) is not suitable for windows and wavelets with multimodal (i.e. having multiple non-negligible peaks) $|\hat{g}(\xi)|$ and $|\hat{\psi}(\xi > 0)|$, respectively. Thus, for such window/wavelet functions it might be very hard to understand where is the "line" separating the time-frequency regions corresponding to different components, because even for the single tone signal $s(t) \sim \cos \nu t$ there will be multiple peaks in the TFR amplitude. As a possible workaround, one can redefine the TFS to be enclosed between the points of the first upward zero crossing of $\nu_{G,W}(\omega, t) - \omega$ at both ends of $\omega_p(t)$, so that $\nu_{G,W}(\omega_\pm(t), t) - \omega_\pm(t) = 0$ and $\partial_\omega \nu_{G,W}(\omega_\pm(t), t) - 1 > 0$. This definition would be suitable even for a very exotic $\hat{g}(\xi)$ and $\hat{\psi}(\xi)$, though it has its own issues. In any case, multimodal windows/wavelets enormously complicate the resultant WFT/WT and are therefore rarely used, so we retain the original intuitive concept of the TFS.

Remark 3.2.2 The signals presented in Fig. 3.2 correspond to a simple case where the components have disjoint supports in frequency, which allows their separation by simple filtration of the signal in the respective frequency bands (e.g. [0.5, 1.5] Hz for the first component in Fig. 3.2a, c, e, g); the components' amplitudes and phases can then be estimated from the corresponding analytic signals (see Sect. 2.1). Whether such an approach will be more accurate and noise-robust than the TFR-based one depends on many factors, such as how well the frequency bands are chosen. In general, however, the supports of different components can overlap in the signal's FT (e.g. see Fig. 2.1), so there might be no filtration alternative. Nevertheless, the aim of Fig. 3.2 is not to present the case where time-frequency analysis tools are most useful, but only to illustrate the concept of the TFS and related issues in a clear and simple way. The same remark also applies to many other figures in this work.

Given its TFS, the parameters of the component can be reconstructed using the inversion formulas (2.11), (2.17), where one should restrict the integration over ω to only the corresponding time-frequency region $[\omega_-(t), \omega_+(t)]$. Such a method will be referred to as *direct reconstruction*. However, in this way one can estimate only the amplitude and phase of the component (through reconstructing its analytic signal), but not its frequency. Nevertheless, the expressions for the latter, as well as for any

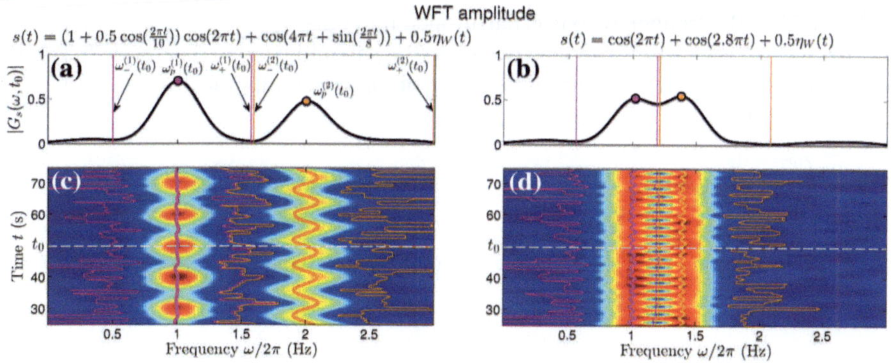

Fig. 3.2 WFT-based examples of the extracted components' time-frequency supports (TFSs) $[\omega_-(t), \omega_+(t)]$ for two different signals, which are defined by the equations above (**a**) and (**b**); in these equations, $\eta_W(t)$ denotes unit-deviation Gaussian white noise. Each signal consists of two components, and in all panels magenta and *orange colors* refer to the first and second component, respectively. **a, b** Snapshots of the WFT amplitudes (*thick black lines*) at time $t_0 = 50$ s, the borders of the components' TFSs at that time (*thin colored lines*) and the associated ridge points $\omega_p(t)$ (*filled colored circles*). **c, d** Full WFT amplitude time-evolutions (t_0, corresponding to snapshots in (**a, b**), are indicated by the *dashed gray lines*), with the components' TFSs $[\omega_-^{(1,2)}(t), \omega_+^{(1,2)}(t)]$ being indicated by *thin colored lines*, and their *ridge curves* $\omega_p^{(1,2)}(t)$ by *thick solid lines* of the same colors. The signals were sampled at 100 Hz for 100s

order time-derivatives of amplitude and phase, can be derived in a similar way to (2.11) and (2.17), as discussed in Sect. 7.3. The direct estimates of the component's parameters are then given as

direct[WFT]:

$$A(t)e^{i\phi(t)} = C_g^{-1}\int_{\omega_-(t)}^{\omega_+(t)} G_s(\omega, t)d\omega,$$

$$\nu(t) = \mathrm{Re}\left[\frac{\int_{\omega_-(t)}^{\omega_+(t)} \omega G_s(\omega, t)d\omega}{\int_{\omega_-(t)}^{\omega_+(t)} G_s(\omega, t)d\omega} - \overline{\omega}_g\right], \; \overline{\omega}_g \equiv C_g^{-1}\frac{1}{2}\int \xi\hat{g}(\xi)d\xi,$$

$$(\overline{\omega}_g = 0 \text{ for symmetric } \hat{g}(\xi), \text{ e.g. Gaussian window}), \tag{3.8}$$

direct[WT]:

$$A(t)e^{i\phi(t)} = C_\psi^{-1}\int_{\omega_-(t)}^{\omega_+(t)} W_s(\omega, t)\frac{d\omega}{\omega},$$

$$\nu(t) = \mathrm{Re}\left[\frac{D_\psi^{-1}\int_{\omega_-(t)}^{\omega_+(t)} \omega W_s(\omega, t)\frac{d\omega}{\omega}}{C_\psi^{-1}\int_{\omega_-(t)}^{\omega_+(t)} W_s(\omega, t)\frac{d\omega}{\omega}}\right], \; D_\psi \equiv \frac{\omega_\psi}{2}\int_0^\infty \frac{1}{\xi}\hat{\psi}^*(\xi)\frac{d\xi}{\xi}.$$

In practice, the frequency axis is partitioned into bins $\omega_k = (k - k_0)\Delta\omega$ (WFT) or $\omega_k/2\pi = 2^{(k-k_0)/n_v}$ (WT), so the formulas (3.8) should be also discretized. Taking into account linear and logarithmic scaling of ω_k, the midpoint discretization rule takes the form

$$\begin{aligned}
&\textbf{WFT:} \int_{\omega_1}^{\omega_2} [\ldots(\omega)]d\omega \to \sum_{k:\omega_k\in[\omega_1,\omega_2]} [\ldots(\omega_k)]\Delta\omega_k \;\; (\text{linear } \omega_k = (k-k_0)\Delta\omega),\\
&\textbf{WT:} \int_{\omega_1}^{\omega_2} [\ldots(\omega)]d\omega \to \sum_{k:\omega_k\in[\omega_1,\omega_2]} \omega_k[\ldots(\omega_k)]\frac{\log 2}{n_v} \;\; \Big(\text{logarithmic } \frac{\omega_k}{2\pi} = 2^{(k-k_0)/n_v}\Big),
\end{aligned} \tag{3.9}$$

where $[\ldots(\omega)]$ stands for any functional of frequency, e.g. in the WT inversion formula (2.17) one has $[\ldots(\omega)] = W_s(\omega,t)/\omega$. If the integrals in (3.8) are discretized according to (3.9), then the related errors in all direct estimates will be $O(\Delta\omega^2)$ (WFT) or $O(n_v^{-2})$ (WT) [10], unless $\hat{g}''(0)$ or $\hat{\psi}''(\omega_\psi)$ are infinite (in which case the errors become $O(\Delta\omega)$ or $O(n_v^{-1})$). These errors are negligible for $\Delta\omega$ and n_v chosen according to (2.34).

Remark 3.2.3 The direct WT-based estimate of instantaneous frequency (3.8) is inapplicable for wavelets characterized by infinite D_ψ (3.8). For the latter to be finite, one needs $|\hat{\psi}(\xi)|$ to decay faster than ξ when $\xi \to 0$, a condition that is not satisfied e.g. for the Morlet wavelet (2.18). In such circumstances one can use some kind of hybrid reconstruction

$$\begin{aligned}
&\textbf{hybrid[WFT]:}\ \nu(t) = \mathrm{Re}\left[\frac{\int_{\omega_-(t)}^{\omega_+(t)} \nu_G(\omega,t)G_s(\omega,t)d\omega}{\int_{\omega_-(t)}^{\omega_+(t)} G_s(\omega,t)d\omega}\right],\ \nu_G(\omega,t) \equiv \partial_t \arg[G_s(\omega,t)],\\
&\textbf{hybrid[WT]:}\ \nu(t) = \mathrm{Re}\left[\frac{\int_{\omega_-(t)}^{\omega_+(t)} \nu_W(\omega,t)W_s(\omega,t)\frac{d\omega}{\omega}}{\int_{\omega_-(t)}^{\omega_+(t)} W_s(\omega,t)\frac{d\omega}{\omega}}\right],\ \nu_W(\omega,t) \equiv \partial_t \arg[W_s(\omega,t)].
\end{aligned} \tag{3.10}$$

Although empirical, it works very well in the majority of cases [10]. However, when both direct and hybrid reconstructions are possible, direct is obviously the more accurate (as it was rigorously derived, see Sect. 7.3); but the difference between the two is usually negligible. Note that, in the case of the WFT, hybrid reconstruction is rarely needed, as direct frequency estimation (3.8) is possible for most of the window functions (being inapplicable only for a very exotic ones with $|\int \xi\hat{g}(\xi)d\xi| = \infty$).

3.2.3 *Difference Between the Two Estimates*

The differences between the direct (3.8) and ridge (3.5) estimates, as well as the related errors, were studied in detail in [10] (see also [7, 11, 13]). It was found that the ridge method is more robust to noise and interference between the components, but that the direct method allows the time variability in the component's parameters at low noise levels to be followed more accurately. Thus, all direct estimates (3.8) by definition give exact values (up to the error of the analytic approximation (2.4)) in the "ideal" case when the TFS contains all the energy of the component and no other contributions. The ridge estimates (3.5), on the other hand, are not exact even in such a perfect case, as they have errors proportional to the strengths of the associated amplitude and frequency modulations. However, direct reconstruction is very susceptible to noise and interference, while ridge reconstruction is more resistant to these complications.

This is illustrated in Fig. 3.3. As can be seen, when the noise is small (Fig. 3.3a, c, e) the direct estimates are preferable, being nearly exact, while ridge reconstruction underestimates the amplitude and frequency variations of the component. The situation changes for higher noise level (Fig. 3.3b, d, f), where the ridge estimates become slightly more preferable (though the relative performance of the two approaches in respect to each other varies with time). Thus, the inaccuracies of both reconstruction methods can be represented as a sum of the theoretical error (the one appearing in the perfect noiseless case) and the error attributable to noise and interference. The former is zero for direct estimates and non-zero for ridge estimates, while the latter is non-zero for both methods, but is smaller in the case of ridge reconstruction [10]. In the case of Fig. 3.3b, d, f the original (theoretical) inaccuracies of the ridge method are largely compensated by its higher noise-robustness, thus making the total errors smaller than for direct reconstruction (at least on the central time interval being shown).

3.2.4 Adaptive Choice of the Reconstruction Method

As discussed in the previous section, each of the two reconstruction methods has its own advantages and drawbacks, so the optimal choice depends on the situation. To select the most appropriate estimates adaptively, one can use the approach proposed in [10], which is discussed below.

Having extracted the ridge curve $\omega_p(t)$ and TFS $[\omega_-(t), \omega_+(t)]$ of the component, its parameters are first reconstructed by both methods (direct and ridge). In this way one obtains the amplitudes $A^{(d,r)}(t)$, phases $\phi^{(d,r)}(t)$ and frequencies $\nu^{(d,r)}(t)$, where "d" and "r" denote direct and ridge estimates, respectively. One then calculates the TFR (using the same parameters as originally) of the signal $s^{(d)}(t) = A^{(d)}(t)\cos\phi^{(d)}(t)$, extracts the ridge curve and TFS from it (taking simple maxima $\omega_p(t) = \operatorname{argmax}_\omega |H_s(\omega, t)|$ is sufficient here), and reconstructs by the direct method the "refined" parameters $\tilde{A}^{(d)}(t)$, $\tilde{\phi}^{(d)}(t)$, $\tilde{\nu}^{(d)}(t)$. The same procedure is performed for the "ridge" signal $s^{(r)}(t) = A^{(r)}(t)\cos\phi^{(r)}(t)$, now using the ridge method to reconstruct the refined estimates. Next, the discrepancies $\varepsilon^{(d,r)}_{A,\phi,\nu}$ between the original and refined estimates are calculated for each method as

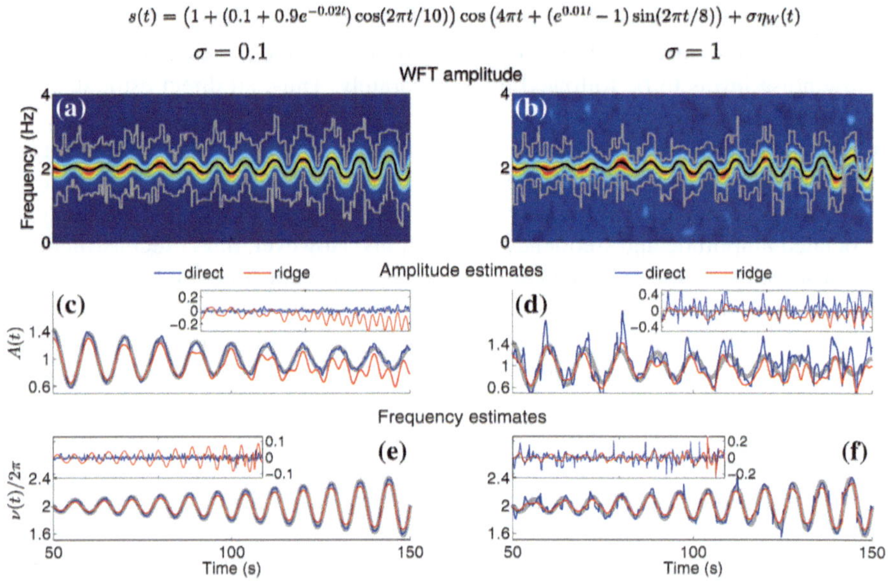

Fig. 3.3 Comparison of the estimates of a component's amplitude and frequency obtained using ridge and direct reconstruction methods in two cases: low noise level (**a**, **c**, **e**) and high noise level (**b**, **d**, **f**). The signal composition is shown by the equation at the *top*, where $\eta_W(t)$ denotes Gaussian white noise of unit deviation. **a**, **b** WFTs of the signals at two noise levels, with the extracted *ridge curves* $\omega_p(t)$ and the associated time-frequency supports $[\omega_-(t), \omega_+(t)]$ being shown by *black* and *gray solid lines*, respectively. **c**, **d** the corresponding direct (*blue*) and ridge (*red*) amplitude estimates compared with the true amplitude (*thick gray background line*); related errors for the same time intervals are shown in the insets. **e**, **f** The corresponding direct (*blue*) and ridge (*red*) frequency estimates compared with the true frequency (*thick gray background line*); related errors for the same time intervals are shown in the insets. Note that the errors in the ridge estimates of amplitude depend largely on the strength of the accompanying frequency modulation, as is clear from (**c**, **e**). The signals were sampled at 50 Hz for 200s, and the same white noise realizations $\eta_W(t)$ were used for both (**a**, **c**, **e**) and (**b**, **d**, **f**)

$$\begin{aligned}
\varepsilon_A^{(d,r)} &\equiv \kappa_A^{(d,r)} \sqrt{\langle(\tilde{A}^{(d,r)}(t) - A^{(d,r)}(t))^2\rangle}, \\
\varepsilon_\phi^{(d,r)} &\equiv \kappa_\phi^{(d,r)} \sqrt{1 - |\langle e^{i(\tilde{\phi}^{(d,r)}(t) - \tilde{\phi}^{(d,r)}(t))}\rangle|^2}, \\
\varepsilon_\nu^{(d,r)} &\equiv \kappa_\nu^{(d,r)} \sqrt{\langle(\tilde{\nu}^{(d,r)}(t) - \nu^{(d,r)}(t))^2\rangle},
\end{aligned} \tag{3.11}$$

where $\kappa_{A,\phi,\nu}^{(d,r)}$ are the coefficients that can be used to tune the performance of the approach (they were found empirically to be $\kappa_{A,\phi,\nu}^{(d)} = \{3, 4, 2\}$, $\kappa_{A,\phi,\nu}^{(r)} = 1$). Obviously, for the exact estimates one has $\varepsilon_{A,\phi,\nu} = 0$, so it is natural to assess the relative performance of the reconstruction methods based on the associated discrepancies (3.11). Therefore, the direct estimate of the amplitude $A^{(d)}(t)$ seems to be preferable to its ridge estimate $A^{(r)}(t)$ if $\varepsilon_A^{(d)} < \varepsilon_A^{(r)}$, and vice versa; the same considerations apply to the phase and frequency estimates.

Despite being empirical, the approach outlined above works very well in practice, selecting the best estimates in the majority of cases. This is illustrated in Fig. 3.4, where the discrepancies (3.11) are shown together with the actual reconstruction errors for each method. As can be seen, the values of $\varepsilon_{A,\phi,\nu}^{(d,r)}$ are proportional to the true errors and allow one to judge reliably about the relative performance of the two reconstruction methods. Thus, for a single tone signal embedded in noise the ridge estimates are always preferred, which is because in this case they do not have theoretical errors (i.e. are exact when the noise is absent) [10]; the criteria based on (3.11) correctly reflects this fact, as can be seen from Fig. 3.4a–c. Next, when amplitude/frequency modulation is present, at low noise levels the direct estimates are preferred, but with increase of the noise strength their inaccuracy grows faster than in the case of ridge reconstruction. Therefore, beyond some threshold noise level (indicated by gray vertical dashed lines in Fig. 3.4) ridge estimates become the more accurate; this threshold and the optimal method in each case can be well recovered from the behavior of the discrepancies (3.11), as is clear from Fig. 3.4d, h, i.

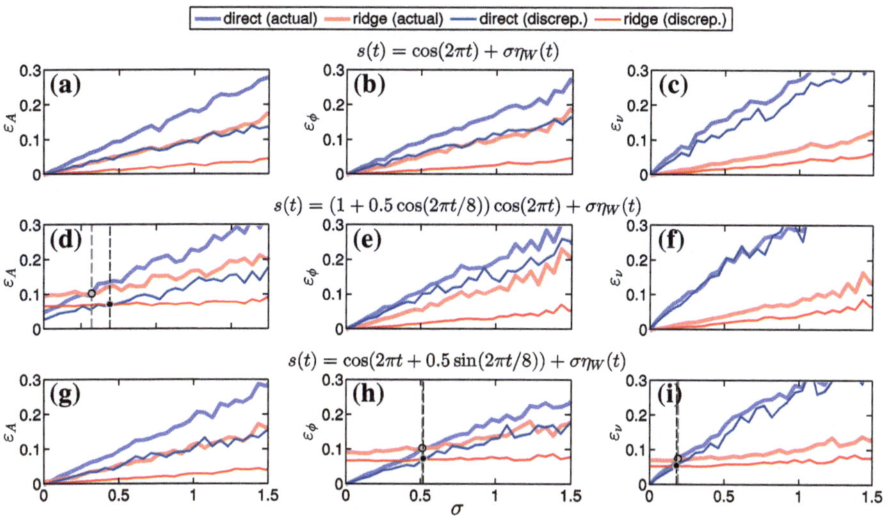

Fig. 3.4 The actual reconstruction errors of the direct and ridge methods (*light-blue* and *light-red lines*, respectively) and the corresponding discrepancies (3.11) (direct—*blue*, ridge—*red*) in their dependence on the noise level σ. The actual inaccuracies are calculated in the same way as (3.11), but using the true parameters instead of the refined ones ($\tilde{A}^{(d,r)}(t) \to A^{(true)}(t)$, $\tilde{\phi}^{(d,r)}(t) \to \phi^{(true)}(t)$, $\tilde{\nu}^{(d,r)}(t) \to \nu^{(true)}(t)$) and setting all $\kappa_{A,\phi,\nu}^{(d,r)} = 1$. **a, d, g** Amplitude reconstruction errors. **b, e, h** Phase reconstruction errors. **c, f, i** Frequency reconstruction errors (for the cyclic frequency $\nu(t)$, in rad/s). The signals associated with each row are given by the equations above the central panels (**b, e, h**), with $\eta_W(t)$ denoting unit-deviation Gaussian white noise; each signal was sampled at 50 Hz for 200 s. Where present, the *gray* (or *black*) points with the corresponding *dashed lines* indicate the intersections between the true errors (or the discrepancies (3.11)) of the direct and ridge methods

Remark 3.2.4 To implement the criterion (3.11), one should compute the two new TFRs, for the direct and ridge signals $s^{(d,r)}(t)$. Obviously, these TFRs do not need to be calculated for all frequencies (which might be computationally expensive), but only for the frequency range $[\omega^{(1)}_{\min}, \omega^{(1)}_{\max}]$ where the original component resides. Given the extracted component's TFS $[\omega_-(t), \omega_+(t)]$, this range can be taken as $[\min \omega_-(t), \max \omega_+(t)]$. However, the maximum of $\omega_+(t)$ might be excessively large (it is in theory infinite e.g. if the signal is represented by a single tone); the same applies to the minimum of $\omega_-(t)$. Therefore, it is better to utilize a narrower support $[\widetilde{\omega}_-(t), \widetilde{\omega}_+(t)]$, defined as the widest region of non-negligible TFR amplitude within the original TFS:

$$\begin{aligned} \widetilde{\omega}_-(t) &: \; |H_s(\omega \in [\omega_-(t), \widetilde{\omega}_-(t)], t)| < \epsilon_s |H_s(\omega_p(t), t)|, \\ \widetilde{\omega}_+(t) &: \; |H_s(\omega \in [\widetilde{\omega}_+(t), \omega_+(t)], t)| < \epsilon_s |H_s(\omega_p(t), t)|, \end{aligned} \tag{3.12}$$

where $H_s(\omega, t)$ stands for the WFT or WT, ϵ_s denotes chosen precision (the default is set to $\epsilon_s = 0.001$), and $\omega_p(t)$ denotes the positions of the associated peaks at each time, as always. If $\widetilde{\omega}_-(t)$ (3.12) does not exist, e.g. if $|H_s(\omega_-(t), t)| \geq \epsilon_s |H_s(\omega_p(t), t)|$, it is taken as $\widetilde{\omega}_-(t) = \omega_-(t)$; the same is done for $\widetilde{\omega}_+(t)$ in a similar case.

With the use of (3.12) the frequency range in which TFR should be calculated (to encompass most of the component's energy) can be estimated as

$$[\omega^{(1)}_{\min}, \omega^{(1)}_{\max}] = \Big[\min\big\{\underset{0.05}{\mathrm{perc}}[\widetilde{\omega}_-(t)], \min[\omega_p(t)]\big\}, \max\big\{\underset{0.95}{\mathrm{perc}}[\widetilde{\omega}_+(t)], \max[\omega_p(t)]\big\}\Big], \tag{3.13}$$

where perc_p denotes the p'th largest percentage value of the argument (with perc_0 and perc_1 corresponding to usual minimum and maximum, respectively). The 95th percentile of $\widetilde{\omega}_+(t)$ is taken in (3.13) instead of the overall maximum because, when noise is present, $\widetilde{\omega}_+(t)$ might sometimes be too large; the same applies to $\widetilde{\omega}_-(t)$.

3.3 Summary

This Chapter has elaborated in detail the extraction of components from the signal's WFT/WT. Thus, to extract the component one should first identify it by finding the corresponding ridge curve (Sect. 3.1), and then reconstruct it using either the ridge (Sect. 3.2.1) or direct (Sect. 3.2.2) method; the choice among the latter two can be made adaptively using the criteria suggested in Sect. 3.2.4. The techniques developed in this Chapter are already very powerful and can be used stand-alone to analyse the properties of the components present in the signal. In the next Chapter, however, they will be combined with some additional procedures to form an even more powerful NMD method.

References

1. A.K. Barros, N. Ohnishi, Heart instantaneous frequency (HIF): an alternative approach to extract heart rate variability. IEEE Trans. Biomed. Eng. **48**(8), 850–855 (2001)
2. R. Bellman, *Dynamic Programming* (Princeton University Press, Princeton NJ, 1957)
3. D.P. Bertsekas, *Dynamic Programming and Optimal Control* (Athena Scientific, Belmont, 1995)
4. R.A. Carmona, W.L. Hwang, B. Torresani, Characterization of signals by the ridges of their wavelet transforms. IEEE Trans. Sig. Proc. **45**(10), 2586–2590 (1997)
5. R.A. Carmona, W.L. Hwang, B. Torresani, Multiridge detection and time-frequency reconstruction. IEEE Trans. Sig. Proc. **47**(2), 480–492 (1999)
6. I. Daubechies, J. Lu, H. Wu, Synchrosqueezed wavelet transforms: an empirical mode decomposition-like tool. Appl. Comput. Harmon. Anal. **30**(2), 243–261 (2011)
7. N. Delprat, B. Escudie, P. Guillemain, R. Kronland-Martinet, P. Tchamitchian, B. Torrésani, Asymptotic wavelet and Gabor analysis: extraction of instantaneous frequencies. IEEE Trans. Inf. Theor. **38**(2), 644–664 (1992)
8. D. Iatsenko, A. Bernjak, T. Stankovski, Y. Shiogai, P.J. Owen-Lynch, P.B.M. Clarkson, P.V.E. McClintock, A. Stefanovska, Evolution of cardio-respiratory interactions with age. Phil. Trans. R. Soc. Lond. A **371**(1997), 20110622 (2013)
9. D. Iatsenko, P.V.E. McClintock, A. Stefanovska, On the extraction of instantaneous frequencies from ridges in time-frequency representations of signals. (2014), http://arxiv.org/abs/1310.7276
10. D. Iatsenko, P.V.E. McClintock, A. Stefanovska, Linear and synchrosqueezed time-frequency representations revisited: overview, standards of use, reconstruction, resolution, concentration, and algorithms. Dig. Sig. Proc. (2015). doi:10.1016/j.dsp.2015.03.004 (in press)
11. J.M. Lilly, S.C. Olhede, On the analytic wavelet transform. IEEE Trans. Inf. Theor. **56**(8), 4135–4156 (2010)
12. M.B. Lotric, A. Stefanovska, D. Stajer, V. Urbancic-Rovan, Spectral components of heart rate variability determined by wavelet analysis. Physiol. Meas. **21**(4), 441 (2000)
13. S. Mallat, *A Wavelet Tour of Signal Processing*, 3rd edn. (Academic Press, Burlington, 2008)
14. G. Thakur, E. Brevdo, N.S. Fuckar, H.-T. Wu, The synchrosqueezing algorithm for time-varying spectral analysis: robustness properties and new paleoclimate applications. Sig. Process. **93**(5), 1079–1094 (2013)

Chapter 4
Nonlinear Mode Decomposition (NMD)

The extraction of the AM/FM components from the signal's WFT/WT has been considered in detail in the previous Chapter, so now it should be clear how to decompose the signal into its constituent components using TFR-based approaches. Thus, having extracted one component, it is then subtracted from the signal, and the procedure is iterated on the residual to extract the next ones. However, it is not clear when to stop this decomposition, i.e. how many components should be extracted. Furthermore, some of the components typically do not correspond to an independent activity, but arise due to the complex waveforms of the particular oscillations to which they are related. This is because real oscillations are rarely purely sinusoidal, but have more complicated shapes as the result of nonlinearities in the generating system and/or the measurement apparatus. For example, the AM/FM component (2.1) raised to the third power $[A(t)\cos\phi(t)]^3 = \frac{3}{4}A^3(t)[\cos\phi(t) + \frac{1}{3}\cos 3\phi(t)]$ already consists of two components, although there is only one meaningful oscillation.

It is therefore better to consider the full *Nonlinear Modes* (NMs), defined as the sum of all components corresponding to the same activity:

$$c(t) = A(t)\mathrm{v}(\phi(t)) = A(t)\sum_h a_h \cos(h\phi(t) + \varphi_h), \tag{4.1}$$

where $\mathrm{v}(\phi(t)) = \mathrm{v}(\phi(t) + 2\pi)$ is some periodic function of phase (also called the "wave-shape function" [1]), which due to its periodicity can always be expanded in the Fourier series (4.1). In what follows, the AM/FM components composing the NM will be referred to as *harmonics*, with the hth harmonic being represented by a term $\sim\cos(h\phi(t) + \varphi_h)$ in (4.1). Without loss of generality, the normalization of $A(t)$ and $\phi(t)$ in (4.1) can be fixed by setting $a_1 = 1, \varphi_1 = 0$. Then the instantaneous phase and frequency of the whole NM are those of its first harmonic (which will be called main, or fundamental), being $\phi(t)$ and $\nu(t) = \phi'(t)$, respectively [1].

Throughout this work, the signal is assumed to have the form (2.5), being composed of the AM/FM components and noise. Obviously, any such signal can be represented as a sum of NMs $c_i(t)$ of the form (4.1), corrupted by some noise $\eta(t)$ (its class will be considered later, in Sect. 4.1.5):

D. Iatsenko, *Nonlinear Mode Decomposition*,
Springer Theses, DOI 10.1007/978-3-319-20016-3_4

$$s(t) = \sum_i c_i(t) + \eta(t). \tag{4.2}$$

The ultimate goal is then to extract all the NMs present in the signal and to find their characteristics, such as the corresponding amplitudes $A(t)$, phases $\phi(t)$ and frequencies $\nu(t)$, as well as the amplitude scaling factors a_h and phase shifts φ_h of the harmonics.

To achieve this goal, the Nonlinear Mode Decomposition (NMD) method is developed and described in the present Chapter. The basic idea and ingredients of the NMD, as well as their implementation, are first considered in Sect. 4.1. Then a few important upgrades are introduced in Sect. 4.2. Finally, the choice of parameters for the method is discussed in Sect. 4.3, and the full procedure is summarized in Sect. 4.4.

4.1 Basic Idea and Ingredients

The main goal of NMD is to decompose a given signal into a set of nonlinear modes (4.1). To do this, four steps are necessary:

(a) Extract the fundamental harmonic of an NM accurately from the signal's TFR.
(b) Find candidates for all its possible harmonics, based on its properties.
(c) Identify the true harmonics (i.e. corresponding to the same mode) among them.
(d) Reconstruct the full NM by summing together all the true harmonics; subtract it from the signal, and iterate the procedure on the residual until a preset stopping criterion is met.

These individual subprocedures are explained in detail in the sections below.

The NMD can be based either on the WFT, or on the WT, although in what follows (Sect. 4.3.2) the choice of the TFR type will be made adaptive. Nevertheless, for now it will be assumed that a particular type has already been selected and all operations are performed based on it. Furthermore, the preferred reconstruction method (direct or ridge, see Sect. 3.2) might differ for different NMD subprocedures, as will be discussed in Sect. 4.3.3; until then, all parameters are assumed to be estimated using some chosen method.

Remark 4.1.1 Before applying NMD the signal should be detrended, which can conveniently be done by subtracting a third order polynomial fit from it (see Sect. 2.4.1). Thus, one cannot decompose the trend, which can be perceived as part of an oscillation of period larger than the time duration of the signal. Moreover, trends, if not removed, might considerably affect the surrogate test developed later in Sect. 4.1.5.

4.1.1 Extraction of the Fundamental Harmonic

Extraction of components from the signal's TFR has been considered in detail in Chap. 3. Thus, applying the procedure of Sect. 3.1, one can identify the ridge curve corresponding to the dominant component, while its associated amplitude/phase/frequency $A^{(1)}(t)$ / $\phi^{(1)}(t)$ / $\nu^{(1)}(t)$ can then be estimated as discussed in Sect. 3.2. The component extracted in this way will generally represent the harmonic of some NM, but not necessarily the fundamental (first) one. For simplicity, however, it will be assumed that this component is the fundamental harmonic; this assumption will be removed in Sect. 4.1.4.

4.1.2 Harmonics: Extracting the Candidates

Given the ridge frequency $\omega_p^{(1)}(t)$ of the first harmonic, the hth harmonic is expected to lie near $h\omega_p^{(1)}(t)$. Therefore, its ridge curve $\omega_p^{(h)}(t)$ can be identified simply as the sequence of peaks which are located in the same time-frequency support (region of unimodal TFR amplitude at each time, see Sect. 3.2.2) as $h\omega_p^{(1)}(t)$; or, in other words, the sequence of peaks nearest to $h\omega_p^{(1)}(t)$ in the direction of increasing TFR amplitude. This is illustrated in Fig. 4.1. Having found $\omega_p^{(h)}(t)$, the parameters of the hth harmonic $A^{(h)}(t)$, $\phi^{(h)}(t)$, $\nu^{(h)}(t)$ can be estimated in the usual way (see Sect. 3.2).

4.1.3 Harmonics: Identifying the True Ones

In general, the procedure of the previous subsection yields what is not necessarily a genuine harmonic, but just a candidate for one. Thus, even if the NM does not contain a particular harmonic, one will still obtain some signal for it, consisting

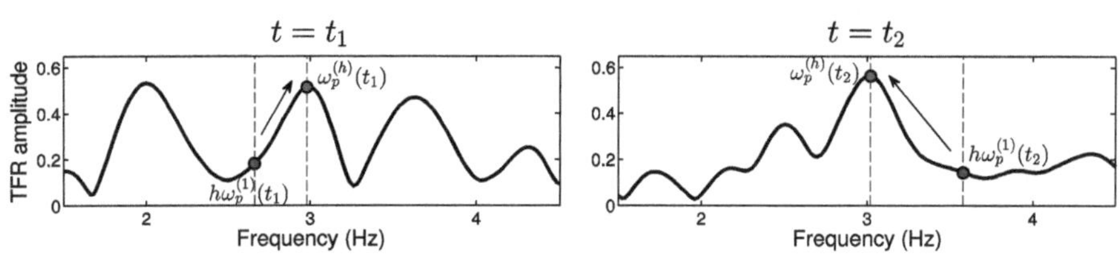

Fig. 4.1 Illustration of the extraction of the hth harmonic ridge curve $\omega_p^{(h)}(t)$ based on the fundamental ridge frequency $\omega_p^{(1)}(t)$. At each time, starting from the expected ridge frequency $h\omega_p^{(1)}(t)$ of the harmonic (*blue points*), one climbs (i.e. follows in the direction of increasing TFR amplitude) to the nearest peak, which is then assigned to $\omega_p^{(h)}(t)$ (*red points*)

of noise or components lying near its expected frequency. Hence, having extracted the hth harmonic candidate, one needs to determine whether it is a true harmonic or not. To tackle this problem, one can use the method of surrogate data [2, 3], testing against the null hypothesis of independence between the first harmonic and the extracted harmonic candidate. Thus, one first selects a measure to quantify the degree of dependence between the dynamics of two harmonics, which is then calculated for the original harmonic and for a collection of surrogates—a specially constructed time-series consistent with the null hypothesis being tested. If the original value of the selected measure lies outside the distribution of its surrogate values, this indicates genuine interdependence and the harmonic is regarded as true; otherwise, it is discarded as false.

The amplitude, phase and frequency dynamics of a true harmonic should depend on those for the fundamental harmonic as $A^{(h)}(t)/A^{(1)}(t) \equiv a_h = \text{const}$, $\phi^{(h)}(t) - h\phi^{(1)}(t) \equiv \varphi_h - h\varphi_1 = \text{const}$, $\nu^{(h)}(t) = h\nu^{(1)}(t)$. One can introduce measures $q^{(h)}_{A,\phi,\nu} \in [0, 1]$ quantifying the degree of consistency with these laws (0—no consistency, 1—full consistency), i.e. the dependence of the parameters of the hth harmonic on the corresponding parameters of the first one:

$$
\begin{aligned}
q_A^{(h)} &\equiv \exp\left[-\frac{\sqrt{\left\langle\left[A^{(h)}(t)\langle A^{(1)}(t)\rangle - A^{(1)}(t)\langle A^{(h)}(t)\rangle\right]^2\right\rangle}}{\langle A^{(1)}(t)A^{(h)}(t)\rangle}\right],\\
q_\phi^{(h)} &\equiv \left|\left\langle \exp\left[i\left(\phi^{(h)}(t) - h\phi^{(1)}(t)\right)\right]\right\rangle\right|, \\
q_\nu^{(h)} &\equiv \exp\left[-\frac{\sqrt{\left\langle\left[\nu^{(h)}(t) - h\nu^{(1)}(t)\right]^2\right\rangle}}{\langle \nu^{(h)}(t)\rangle}\right].
\end{aligned}
\tag{4.3}
$$

The overall measure of interdependence between the harmonics can then be taken as

$$
\rho^{(h)}(w_A, w_\phi, w_\nu) = \left(q_A^{(h)}\right)^{w_A}\left(q_\phi^{(h)}\right)^{w_\phi}\left(q_\nu^{(h)}\right)^{w_\nu}, \tag{4.4}
$$

with the parameters $w_{A,\phi,\nu}$ giving weights to each of the consistencies $q^{(h)}_{A,\phi,\nu}$. The default is set to $\rho^{(h)} \equiv \rho^{(h)}(1, 1, 0)$, with equal weights for the amplitude and phase consistencies, and no weight for the frequency consistency. The latter is because the procedure of harmonic extraction (see Sect. 4.1.2), being based on the instantaneous frequency of the first harmonic, in itself introduces a dependence of $\nu^{(h)}(t)$ on $\nu^{(1)}(t)$, so it is better not to base any conclusions on it.

Ideally, for true harmonics one should have $q^{(h)}_{A,\phi,\nu} = 1$, but noise, the finite frequency and time resolutions, and the interference with other components all introduce errors. In reality, therefore, the consistencies will be smaller than unity even for true harmonics. Hence one cannot identify harmonics based only on the value of $\rho^{(h)}$ but also needs to perform the surrogate test. For the latter one can utilize the idea of the time-shifted surrogates [4, 5], using as a first harmonic surrogate its time-shifted version and as the other harmonic surrogate the corresponding candidate harmonic extracted from the time-shifted TFR. Such time-delay destroys any temporal correlations between the signals while preserving all their other properties, thus creating surrogates consistent with the null hypothesis of independence.

Given the maximal time-shift (in samples) M, the surrogate parameters for the first harmonic are taken as its original parameters shifted $\Delta T_d/2$ backward in time

$$\{A_d^{(1)}(\tau), \phi_d^{(1)}(\tau), \nu_d^{(1)}(\tau)\} = \{A^{(1)}(\tau - \Delta T_d/2), \phi^{(1)}(\tau - \Delta T_d/2), \nu^{(1)}(\tau - \Delta T_d/2)\},$$
$$\tau = \{t_{n=1+M/2,...,N-M/2}\},\ \Delta T_{d=1,...,N_d} = M(1 - 2d/N_d)/2f_s, \tag{4.5}$$

where N_d is the number of surrogates, f_s is the signal sampling frequency and N is its total length in samples (note, that the length of the surrogate time series is smaller than the length of the original signal, being $N - M$). Using $\nu_d^{(1)}(\tau)$ (4.5) as a reference profile, the surrogate hth harmonic and its parameters $A_d^{(h)}(\tau), \phi_d^{(h)}(\tau), \nu_d^{(h)}(\tau)$ are extracted in the same way as described in Sect. 4.1.2, but from the signal's TFR shifted $\Delta T_d/2$ forward in time ($G(\omega, \tau + \Delta T_d/2)$ or $W(\omega, \tau + \Delta T_d/2)$).

The extraction of the corresponding curves for all surrogates can greatly be accelerated by initial preprocessing of the TFR, constructing its "skeleton" [6] at the beginning and then utilizing it for each surrogate. Thus, at each time t one breaks the TFR into the time-frequency supports (regions of unimodal TFR amplitude) $[\omega_{-,m}(t), \omega_{+,m}(t)], m = 1, 2, ..., N_p(t)$, and from them reconstructs by (3.5) or (3.8) the corresponding amplitudes $\overline{A}_m(t)$, phases $\overline{\phi}_m(t)$ and frequencies $\overline{\nu}_m(t)$ (which form the TFR skeleton). Then the surrogate parameters are taken as those corresponding to indices $m_d(t)$ of the supports in which the expected harmonic frequency lies:

$$m_d(\tau):\ h\nu_d^{(1)}(\tau) \equiv h\nu^{(1)}(\tau - \Delta T_d/2) \in [\omega_{-,m_d(\tau)}(\tau + \Delta T_d/2), \omega_{+,m_d(\tau)}(\tau + \Delta T_d/2)]$$
$$\{A_d^{(h)}(\tau), \phi_d^{(h)}(\tau), \nu_d^{(h)}(\tau)\} = \{\overline{A}_{m_d(\tau)}(\tau), \overline{\phi}_{m_d(\tau)}(\tau), \overline{\nu}_{m_d(\tau)}(\tau)\}. \tag{4.6}$$

Remark 4.1.2 Instead of re-extracting $A_d^{(h)}(\tau), \phi_d^{(h)}(\tau), \nu_d^{(h)}(\tau)$ for each time-delay, one can take the surrogate parameters for the hth harmonic simply by shifting its original $A^{(h)}(t), \phi^{(h)}(t), \nu^{(h)}(t)$ forward in time by $\Delta T_d/2$, in the same way as is done for the fundamental harmonic (4.5). In fact, such an approach gives almost the same results and at the same time is faster. However, as mentioned before, the original procedure of harmonic extraction introduces a correlation between $\nu^{(h)}(t)$ and $\nu^{(1)}(t)$, and can therefore introduce dependence between other parameters, such as phases. Therefore, to be rigorous, the full extraction procedure should be repeated for each surrogate, thus automatically taking into account possible bias of this kind and producing more reliable estimates of the significance.

Summarizing, to perform the surrogate test one calculates the amplitude-phase consistencies $\rho_{d=1,...,N_d}^{(h)}(1, 1, 0)$ (4.4) for the surrogate parameters (4.5), (4.6) and compares them with the value $\rho_0^{(h)}(1, 1, 0)$ calculated in the same way but for the zero time shift $\Delta T_0 = 0$. The probability measure (although mathematically not the true probability) that the extracted hth harmonic is a true harmonic of the main one is then quantified by the significance of the surrogate test, i.e. by the relative part of surrogates for which $\rho_d^{(h)} < \rho_0^{(h)}$. For example, if one found 1000 surrogate amplitude-phase consistencies $\rho_{d=1,\ldots,1000}^{(h)}$ and 792 of them are smaller than the original value $\rho_0^{(h)}$, then the rough probability that the extracted harmonic candidate represents a true harmonic is 79.2 %. Following the standard convention [2, 3], in what follows the harmonic is regarded as true if the probability calculated in this way is $\geq$95 %. By default, the number of surrogates is set to $N_d = 100$, while the maximum time shift equals $M = N/4$, i.e. one quarter of the signal's length, so that the surrogates are each of length $N - M = 3N/4$.

Note that the significance of the surrogate test does not depend on the magnitude of $\rho^{(h)}$ (4.4). Thus, there might be an independent component located at the frequency of a possible harmonic, so that the amplitude-phase consistency would be high but, because it does not fully adjust its amplitude and phase to that of the fundamental harmonic, the surrogate test will not reject the null hypothesis of independence. Thus, the possibility of picking up a spurious component that is nearby in frequency is largely eliminated.

Remark 4.1.3 Since many harmonic candidates $h = 2, 3, \ldots$ are tested for being true, it is natural to expect that sometimes one might encounter false positives, i.e. the surrogate test will regard as true a harmonic which is actually false. Thus, for some noise realizations the parameters of the TFR around the expected harmonic frequency might indeed appear to be correlated with the parameters of the fundamental harmonic. However, in such cases the extracted harmonic candidate will usually have quite a small consistency $\rho^{(h)}$ (4.4). To reduce the probability of false positives, therefore, it is reasonable to introduce some threshold $\rho_{\min}$ and to regard the harmonic as true only if it both passes the surrogate test and at the same time has $\rho^{(h)} \geq \rho_{\min}$. Empirically, this threshold is set to

$$\rho_{\min}(w_A, w_\phi, w_\nu) = 0.5^{w_A + w_\phi}, \tag{4.7}$$

where w_A, w_ϕ, w_ν are the weightings used in (4.4) (as mentioned, the default is $w_A = w_\phi = 1,\ w_\nu = 0$); the value (4.7) was found to be quite universal and to work well in most cases. Note, that for a true harmonic one can also have $\rho^{(h)} < \rho_{\min}$, but this will usually mean that it is badly damaged by noise or other influences and cannot be recovered without large errors.

4.1.4 Harmonics: Practical Issues

Extracting in order To improve the accuracy of reconstruction, each harmonic which is identified as true should be subtracted from the signal before extracting and testing the next one. This decreases the errors related to interference between the harmonics and makes all procedures more accurate. The same consideration applies to the first harmonic which, after being found by the methods described in Sect. 4.1.1, should be subtracted from the signal before extraction of any of the other harmonics.

How many harmonics to extract? Clearly, the maximum number of harmonics one can extract in principle is $h_{\max} = (f_s/2)/\langle \nu^{(1)}(t)/2\pi \rangle$, where f_s is the sampling frequency of the signal (so that $f_s/2$ is the maximum achievable, being the Nyquist frequency) and $\nu^{(1)}(t)$ denotes the extracted instantaneous frequency of the first harmonic. However, checking all harmonics might be computationally very expensive, and is often not needed. In practice, it is better to stop the search after some chosen number S of sequential harmonics has been identified as false, making it likely that there are no more true harmonics. This number is chosen to be $S = 3$ by default.

What if the extracted component is not the first harmonic? Although intuitively the first harmonic should have the highest amplitude (and will therefore be extracted as the dominant curve), for some complicated waveforms this might be untrue. Therefore, before extracting the higher harmonics, it should first be ensured that one starts from the first harmonic. To do this, the same procedure described for harmonic extraction and identification can be applied, but in the reverse direction, i.e. using $h = 1/2, 1/3, 1/4, \ldots$. Then if some of these "fractional" harmonics are identified as true, the one with the smallest frequency is assigned as the new fundamental harmonic. The minimum h one can go for can be set as $h_{\min} = (1/T)/\langle \nu^{(1)}(t)/2\pi \rangle$, although the statistics will be bad already for $h < 5h_{\min}$, when the related oscillation has less than 5 cycles over the whole signal time-duration (see Sect. 2.4.4). Nevertheless, it is better to stop after $S_0 = 3$ consecutive $1/n$ harmonics have been identified as false, in the same manner as is done for the usual harmonics.

In which frequency range to calculate TFRs for harmonics? To extract a particular harmonic candidate, one does not need to calculate the TFR over the whole frequency range, which would be computationally expensive. Rather, one can calculate it only in the range where the possible harmonic is expected to lie. As discussed in [7], the maximum frequency range $[\omega^{(h)}_{\min}, \omega^{(h)}_{\max}]$ where the hth harmonic is concentrated in the TFR can be estimated as

$$\begin{aligned}
&\overline{\omega}^{(h)} \equiv h\langle \omega_p^{(1)}(t)\rangle, \quad \Delta\omega^{(h)}_{\min,\max} \equiv \max(1,h)\big[\omega^{(1)}_{\min,\max} - \langle \omega_p^{(1)}(t)\rangle\big], \\
&\textbf{WFT: } \omega^{(h)}_{\min,\max} = \overline{\omega}^{(h)} + \Delta\omega^{(h)}_{\min,\max} \max\left(1, \frac{\min(1,h) f_0^{(1)}}{h f_0^{(h)}}\right), \\
&\textbf{WT: } \omega^{(h)}_{\min,\max} = \overline{\omega}^{(h)} \left(\frac{\overline{\omega}^{(h)} + \Delta\omega^{(h)}_{\min,\max}}{\overline{\omega}^{(h)}}\right)^{\max(1,\min(1,h) f_0^{(1)}/f_0^{(h)})},
\end{aligned} \tag{4.8}$$

where $\omega_p^{(1)}(t)$ is the ridge curve for the first harmonic (that is assumed to have been extracted already), $f_0^{(h)}$ is the resolution parameter of the WFT/WT from which the hth harmonic is extracted (it might be adjusted, see Sect. 4.2.1 below), and $[\omega^{(1)}_{\min}, \omega^{(1)}_{\max}]$ denotes frequency range of the first harmonic in the TFR (it can be estimated by (3.13)).

Remark 4.1.4 Note that the formula (4.8) was derived [7] within the assumption that the window (wavelet) function used for the WFT (WT) obeys $\hat{g}[f_0^{(h)}](\xi) \sim \hat{g}[f_0^{(1)}](\xi f_0^{(h)}/f_0^{(1)})$ ($\hat{\psi}[f_0^{(h)}](\xi) \sim \hat{\psi}[f_0^{(1)}](\xi^{f_0^{(h)}/f_0^{(1)}})$), which holds for the Gaussian window (2.12) (lognormal wavelet (2.19)).

4.1.5 Stopping Criterion

Once all the harmonics are identified and reconstructed, they are summed up into the NM, which is then subtracted from the signal and the procedure is repeated on the residual. The natural question arises, therefore, of how many nonlinear modes to extract, i.e. when to stop the decomposition. Obviously, decomposition of any noise (white, Brownian, any other correlated or not) does not make much sense, so the reasonable goal is to extract all oscillatory components present in the signal and leave the noise and trends as the residual. Therefore, after extraction of each NM one needs to decide whether what is left contains any more meaningful oscillations, in which case one continues the decomposition, or whether it just represents noise, in which case one should stop. The problem is thus reduced to distinguishing deterministic from random dynamics.

To solve it, one can use the surrogate test against the null hypothesis of linear noise [2, 3], which includes white and colored noises (e.g. Brownian). The surrogates for this task, called FT surrogates, are constructed by taking the inverse Fourier transform of the signal's FT with randomized phases of the Fourier coefficients: $s_s(t) = (2\pi)^{-1} \int [\hat{s}(\xi) e^{i\varphi_s(\xi)}] e^{i\xi t} d\xi$, where $\varphi_s(\xi) = -\varphi_s(-\xi)$ denote the phases taken at random uniformly on $[0, 2\pi]$ for each frequency $\xi > 0$. The reason for this is that any linear noise (an ARMA process $x(t_n) = a_0 + b_0 \eta_W(t_n) + \sum_{p=1}^{P} a_p x(t_{n-p}) + \sum_{m=1}^{M} b_m \eta_W(t_{n-p})$, where $\eta_W(t_n)$ denotes Gaussian white noise of unit variance) is characterized only by the amplitudes of the Fourier coefficients. Randomization of the Fourier phases preserves the power spectrum, so that the surrogate time series will represent another realization of the same random process if the original time series is noise, thus being consistent with the tested null hypothesis. On the other hand, if some meaningful NMs are present in the signal, the randomization of the phases will destroy the particular phase relationships responsible for the amplitude and frequency modulations, making their behavior less deterministic.

One now needs to select the discriminating statistics, which is calculated for the original signal and the surrogates, so that the null hypothesis of linear noise is accepted if the original value lies within the surrogate values and rejected otherwise. The commonly used statistics involve one of the generalized Renyi dimensions [8–10], with the correlation dimension calculated by the Grassberger-Procaccia approach [11–13] remaining the most popular choice. However, it turns out that the surrogate test based on such measures is extremely sensitive to noise, prohibiting their use in the NMD which is intended to be noise-robust.

Therefore, another discriminating statistics should be devised. There are virtually no restrictions on its choice [14], so that almost any measure can be used. The only question to be asked is how powerful it is, i.e. how good in distinguishing the deterministic dynamics from noise. Given that NMD is based on the WFT/WT, it is reasonable to select statistics based on the properties of the time-frequency representation obtained. Namely, since at the first step of the NMD procedure one extracts the component from the signal's TFR (see Sect. 4.1.1), the discriminating statistics can be based on the properties of the components extracted (in the same way) from the original signal and from its FT surrogates. Thus, if the original component is true (and not just formed from noise peaks picked in the time-frequency plane), then it is expected to have more deterministic amplitude and frequency modulation than the surrogate components, which should be more stochastic; otherwise, there will be no difference.

The degree of order in the extracted amplitude $A(t)$ and frequency $\nu(t)$ can be quantified by their spectral entropies $Q[\hat{A}(\xi)]$ and $Q[\hat{\nu}(\xi)]$, respectively, so that the discriminating statistics D for the surrogate test can be taken as their combination

$$D(\alpha_A, \alpha_\nu) \equiv \alpha_A Q[\hat{A}(\xi)] + \alpha_\nu Q[\hat{\nu}(\xi)],$$
$$Q[f(x)] \equiv -\int \frac{|f(x)|^2}{\int |f(x)|^2 dx} \log \frac{|f(x)|^2}{\int |f(x)|^2 dx} dx. \tag{4.9}$$

Note that, in practice, due to the finite sampling frequency f_s and sample length N of real signals, the integrals over frequency ξ in $Q[\hat{A}(\xi)]$, $Q[\hat{\nu}(\xi)]$ (4.9) are discretized into sums over the discrete FT frequencies $\xi_n/2\pi = (n/N - 1/2)f_s$, $n = 1, ..., N$.

In the present context, the statistics $D(\alpha_A, \alpha_\nu)$ (4.9) appears to be more meaningful and much more powerful than other choices (e.g. the popular correlation dimension [11–13]). This statistics is directly related to the quality of the representation of component in the signal's TFR, so that the significance of the surrogate test based on it reflects the proportion of the "deterministic" part in the extracted amplitude and frequency dynamics. Thus, if the residual does not pass the surrogate test (null hypothesis is not rejected), this might mean either that the residual is indeed noise, or that the component simply cannot be reliably extracted from the TFR (e.g. because the resolution characteristics of the latter are not appropriate to represent reliably the related amplitude/frequency modulation and/or to segregate the component from noise).

The power of $D(\alpha_A, \alpha_\nu)$, i.e. its ability to distinguish deterministic from random dynamics, depends strongly on the complexity of the component's amplitude and frequency modulations: the lower the original spectral entropies $Q[\hat{A}(\xi)]$, $Q[\hat{\nu}(\xi)]$ are, the more powerful is the test. However, even in the (quite unrealistic) case when the signal contains a meaningful component without any amplitude or frequency modulation, i.e. a pure tone $A\cos\nu t$, due to numerical issues [15, 16] the surrogate test will still be quite powerful in rejecting the null hypothesis (unless this tone has an integer number of cycles over the time-length of the signal). The power of the test is also inversely proportional to the spread of $\hat{A}(\xi)$, $\hat{\nu}(\xi)$: the more concentrated they are, the narrower the frequency band that the component $A(t)\cos\phi(t)$ occupies, so that the less the noise power that is contained in it.

As to the choice of α_A, α_ν in (4.9), it turns out that the powers of $D(1,0) = Q[\hat{A}(\xi)]$ and $D(0,1) = Q[\hat{\nu}(\xi)]$ are often inversely proportional to the strengths of the amplitude and frequency modulation, respectively. Thus, $D(1,0)$ is preferable for components with relatively small amplitude variability and considerable frequency variability, while $D(0,1)$ is better otherwise. Therefore, it is reasonable to perform three tests, using $D(1,0)$, $D(0,1)$ and $D(1,1)$ as a discriminating statistics, and then select the significance as the maximum among them. It remains to be established, however, whether some better statistics not having the drawback mentioned can be found.

Summarizing, one first extracts the components from the TFR of the original signal and computes the corresponding $D_0(\alpha_A, \alpha_\nu)$ (4.9); then one creates N_s FT surrogates of the signal, for each of them calculates the corresponding TFR, extracts the component from it and computes the respective $D_{s=1,\ldots,N_s}(\alpha_A, \alpha_\nu)$. The default surrogate number and significance level are set to $N_s = 40$ and 95 %, respectively, so that the tested null hypothesis of noise is rejected if the number of surrogates with $D_s > D_0$ is equal or higher than $0.95 \times 40 = 38$. The test is performed for three different (α_A, α_ν) in (4.9), using $D(1,1)$, $D(0,1)$ and $D(1,0)$ as a discriminating statistic; if at least for one of them the null hypothesis is rejected, the signal is regarded as not noise and the decomposition continues.

Remark 4.1.5 Similarly to the case of harmonics (see Sect. 4.1.4), it is not necessary to calculate the WFT/WT of each surrogate over the whole frequency range. Instead, it is sufficient (and much faster) to calculate the surrogate TFRs only for the frequency range of the original component, which can be estimated by (3.13). Additionally, to guarantee that the boundary distortions are of the same nature in the original TFR and those for the surrogates, it is recommended to use padding with zeros (see Sect. 2.4.3); note that in all other cases the more accurate (but at the same time more computationally expensive) predictive padding is used. To be fully consistent, one should also recalculate the signal's TFR using the same padding (with zeros) and frequency range as for surrogates, extract the component from it and estimate the original discriminating statistics D_0 based on the reconstructed parameters of this updated component.

Remark 4.1.6 Instead of the FT surrogates used here, one can alternatively utilize the more advanced IAAFT surrogates [17], which are devoted to testing against the null hypothesis of an invertibly rescaled linear stochastic process (e.g. Brownian noise taken to the third power). These surrogates preserve both the power spectrum and, as much as possible, the distribution of values of the original signal. However, as can be seen from (2.8) and (2.13), the WFT and WT do not explicitly take into account the distribution of signal values, but only its FT. Therefore, the simpler and easier-to-compute FT surrogates seem to be the more natural choice in the present case, and the test developed above seems to work well even for rescaled (either invertibly or non-invertibly), non-Gaussian and non-stationary linear noise. In fact, the outcome of the test seems to be determined primarily by the "goodness" of the component representation in the TFR, which is affected only by the noise power and its distribution in the time-frequency region around the component's instantaneous frequency.

4.2 Improvements

The NMD as described in the previous section already represents a very powerful decomposition tool. However, it may be made even better with the help of the upgrades outlined below, although at the expense of greatly increased computational cost for some of them (though it still remains $O(N \log N)$).

4.2.1 Adaptive Representation of the Harmonics

Even if the first harmonic of some NM is well represented in the TFR and can be accurately extracted and reconstructed, it does not mean that the same applies to all the other harmonics too. For example, harmonics of an NM with only amplitude modulation require the same time and frequency resolution, so that the WFT can represent them all well, while for the WT, where time and frequency resolution scale with frequency, one will need to adjust the resolution parameter f_0 for each harmonic. Thus, consider an NM with simple sinusoidal amplitude modulation:

$$\begin{aligned} s(t) =&(1 + r_a \cos(\nu_a t + \varphi_a)) \sum_{h=1}^{\infty} a_h \cos(h\nu t + \varphi_h) \equiv \sum_{h=1}^{\infty} x^{(h)}(t) \\ \Rightarrow x^{(h)}(t) &\equiv a_h(1 + r_a \cos(\nu_a t + \varphi_a)) \cos(h\nu t + \varphi_h) \\ &= a_h\Big[\cos(h\nu t + \varphi_h) + \frac{r_a}{2} \cos((h\nu + \nu_a)t + (\varphi_h + \varphi_a)) \\ &+ \frac{r_a}{2} \cos((h\nu - \nu_a)t + (\varphi_h - \varphi_a))\Big]. \end{aligned} \tag{4.10}$$

Clearly, all harmonics have the same amplitude variability (in relative terms), so that the time resolution of the TFR should also be the same for them. This is reflected in the fact that each harmonic $x^{(h)}(t)$ is composed of three Fourier terms, which all have identical amplitude ratios, frequency differences and phase relationships for each h. Furthermore, the frequency distance between two consecutive harmonics remains the same, meaning that the frequency resolution also should not be changed for harmonics.

Therefore, the WFT, having constant time and frequency resolution, will be a perfect match for this case. This means, first, that if the amplitude modulation of the first harmonic is represented reliably in the WFT, then the same will also apply

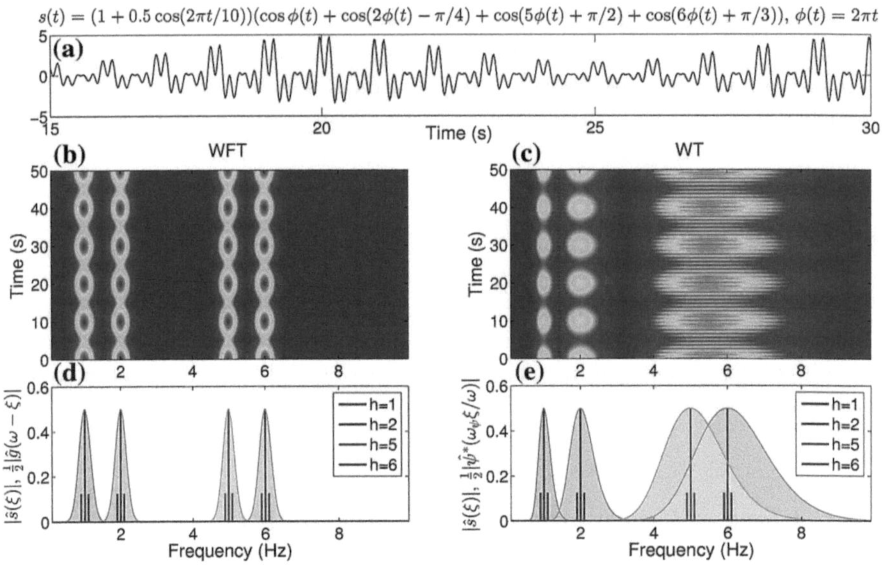

Fig. 4.2 **a** Nonlinear mode with amplitude modulation, as specified by the equation for $s(t)$ above the *box*. **b**, **c** The corresponding WFT and WT amplitudes, respectively. **d**, **e** The signal's FT, integrated over one frequency bin (vertical lines), with the parts responsible for different harmonics shown in different colors; the shaded areas show the absolute values of the window functions $\hat{g}(\omega - \xi)$ or wavelet functions $\hat{\psi}^*(\omega_\psi \xi/\omega)$ centered at the mean frequencies of the harmonics $\omega = 2\pi h$ and rescaled to half of the harmonics' mean amplitudes. The signal was sampled at 50 Hz for 50 s

to all other harmonics and, secondly, that if two first harmonics do not interfere in the WFT, then any two harmonics will also be well-separated. For the WT, the former is also true, as the time resolution, i.e. the ability to reflect changes in time, increases with frequency for this type of TFR. However, the frequency resolution of the WT progressively worsens with the increase of frequency, so that the higher the harmonics are, the harder it is to resolve them.

This is illustrated in Fig. 4.2, where from (b) and (c) it is clear that, for NM without frequency modulation, all harmonics can be well represented in the WFT, but for the WT higher harmonics begin to interfere. This issue is explained schematically in Fig. 4.2d, e. Thus, both the WFT and WT can be seen as convolutions in the frequency domain of the signal with a window $\hat{g}(\omega - \xi)$ and wavelet $\hat{\psi}^*(\omega_\psi \xi/\omega)$, as seen from (2.8) and (2.13). Figures 4.2d, e show the signal's discrete FT $\hat{s}(\xi)$ together with the (rescaled and centered at the mean frequencies of the harmonics $\omega/2\pi = 1, 2, 5, 6$ Hz) Gaussian window (2.12) and lognormal wavelet (2.19) FTs $\hat{g}(\omega - \xi)$ and $\hat{\psi}^*(\omega_\psi \xi/\omega)$, with which $\hat{s}(\xi)$ is convoluted while constructing the WFT and WT, respectively.

Roughly speaking, the quality of the representation of time-variability (the amplitude modulation in the present case) for each harmonic can be estimated based on the proportion of the associated Fourier coefficients lying below the shown window/wavelet FTs. For both the WFT and WT, each harmonic has three Fourier coefficients (4.10), all of which lie in the correspondingly shaded areas, meaning that the time resolution in each case is sufficient to represent amplitude modulation appropriately. The degree of interference between the harmonics in the WFT and WT can be estimated from the area of overlap between the corresponding window/wavelet FTs. For the WFT (Fig. 4.2d) the interference between harmonics does not depend on their frequency, while for the WT (Fig. 4.2e) it increases for the higher harmonics, which results in the fifth and sixth harmonics being represented as a single component (Fig. 4.2c). This is because the minimum frequency difference between two harmonics is equal to the frequency of the first harmonic, thus being defined on a linear scale (which is natural for the WFT), but not on a logarithmic one (natural for the WT).

The situation changes when there is frequency modulation. In this case neither the WFT nor the WT provides optimal representation of the harmonics; in addition, it might be in principle not possible to reliably resolve some harmonics with time-frequency analysis methods. Thus, consider an NM with simple sinusoidal frequency modulation

$$\begin{aligned} s(t) &= \sum_{h=1}^{\infty} a_h \cos(h\phi(t) + \varphi_h) \equiv \sum_{h=1}^{\infty} x^{(h)}(t), \;\; \phi(t) = \nu t + r_b \sin(\nu_b t + \varphi_b) \\ \Rightarrow x^{(h)}(t) &= a_h \cos(h\nu t + h r_b \sin(\nu_b t + \varphi_b) + \varphi_h) \\ &= a_h \mathrm{Re}\Big[\sum_{n=-\infty}^{\infty} J_n(h r_b) e^{i(\varphi_h + n\varphi_b)} e^{i(h\nu + n\nu_b)t} \Big], \end{aligned} \tag{4.11}$$

where the formula $e^{ia\sin\phi} = \sum_{n=-\infty}^{\infty} J_n(a)e^{in\phi}$ has been used. Due to properties of the Bessel functions $J_n(a)$, after some $|n|$ all terms in (4.11) become negligible. Therefore, in practice, one can restrict the summation to $|n| \leq n_J(ha)$, with the maximum non-negligible order n_J being determined as

$$n_J(a) : \frac{\sum_{n:\,|n|>n_J(a)} |J_n(a)|^2}{\sum_{n=-\infty}^{\infty} |J_n(a)|^2} < \epsilon_p, \tag{4.12}$$

where ϵ_p denotes a chosen accuracy threshold. The value of $n_J(a)$ increases with a, being (for $\epsilon_p = 0.001$): $n_J(0 \lesssim a \lesssim 0.3) = 2$, $n_J(0.3 \lesssim a \lesssim 0.65) = 3$, $n_J(0.65 \lesssim a \lesssim 1.13) = 4$ etc. As a result, the higher the harmonic, the larger the frequency range it occupies, i.e. the larger is the number of non-negligible terms in (4.11).

Consequently, to reflect reliably the frequency modulation of the higher harmonics, one needs higher time resolution for them, a requirement that is satisfied by the WT, but not by the WFT. This issue is illustrated in Fig. 4.3, where it can be seen that the WT can represent reliably both the first and third harmonics, while the WFT reflects appropriately only the first one. However, the increased time resolution of the WT is provided at the expense of decreased frequency resolution, leading to stronger

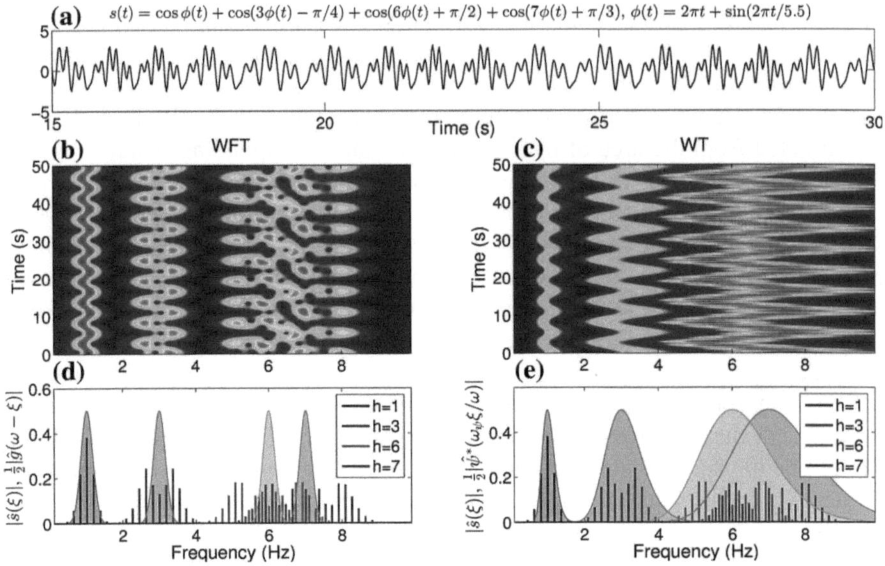

Fig. 4.3 **a** Nonlinear Mode with frequency modulation, as specified by the equation for $s(t)$ above the *box*. **b**, **c** The corresponding WFT and WT amplitudes, respectively. **d**, **e** The signal's FT, integrated over one frequency bin (*vertical lines*), with the parts responsible for different harmonics shown in different colors; the shaded areas show the absolute values of the window functions $\hat{g}(\omega-\xi)$ or wavelet functions $\hat{\psi}^*(\omega_\psi\xi/\omega)$ centered at the mean frequencies of the harmonics $\omega = 2\pi h$ and rescaled to half of the harmonics' mean amplitudes. The signal was sampled at 50 Hz for 55 s

interference between harmonics, as seen from the previous case (Fig. 4.2). Figure 4.3 also shows, that in some cases it might in principle be impossible to represent reliably two harmonics in the TFR. Thus, as can be seen from Fig. 4.3d, e, the FTs $\hat{x}^{(h)}(\xi)$ of the sixth and seventh harmonics are "entangled", i.e. the frequency regions in which they are contained overlap. Therefore, unless specifically designed for this case, any window/wavelet function which picks the Fourier coefficients of one harmonic will inevitably also pick those corresponding to the other one too.

Summarizing, for the general case where both amplitude and frequency modulation are present in the NM, accurate representation of higher harmonics requires higher time resolution, but the same frequency resolution. However, an increase in time resolution inevitably leads to a decrease of frequency resolution, so usually one needs to search for some compromise. Consequently, it is often the case that neither the WFT nor the WT with a constant resolution parameter f_0 can represent all harmonics reliably.

To tackle this problem, one can adaptively adjust f_0 for each harmonic individually. Assuming that the extracted first harmonic is reconstructed well, the quality of the representation of the hth harmonic can be assessed through its consistency $\rho^{(h)}$ (4.4) with the first one. Therefore, the optimal resolution parameter for the hth harmonic, $f_0^{(h)}$, can be selected as that for which $\rho^{(h)}$ (4.4) achieves its maximum and at the same time the harmonic passes the surrogate test, i.e. is identified as a true harmonic.

The remaining question is the region within which to search for $f_0^{(h)}$. To find it, one first needs to understand in what frequency band $[\omega_f^{(h)} - \Delta\omega_f^{(h)}/2, \omega_f^{(h)} + \Delta\omega_f^{(h)}/2]$ the hth harmonic FT $\hat{x}^{(h)}(\xi)$ is concentrated, given the frequency band of the first harmonic. In general, $\omega_f^{(h)}$ and $\Delta\omega_f^{(h)}$ can be defined as

$$\omega_f^{(h)}, \Delta\omega_f^{(h)} : \int\limits_0^{\omega_f^{(h)}-\Delta\omega_f^{(h)}/2} |\hat{x}^{(h)}(\xi)|^2 \frac{d\xi}{2\pi} = \frac{\epsilon_p}{2} \frac{E_{tot}^{(h)}}{2}, \int\limits_{\omega_f^{(h)}+\Delta\omega_f^{(h)}/2}^{\infty} |\hat{x}^{(h)}(\xi)|^2 \frac{d\xi}{2\pi} = \frac{\epsilon_p}{2} \frac{E_{tot}^{(h)}}{2}, \tag{4.13}$$

where $E_{tot}^{(h)} \equiv (2\pi)^{-1} \int |\hat{x}^{(h)}(\xi)|^2 d\xi = \int |x^{(h)}(t)|^2 dt$ is the total energy of the harmonic, and ϵ_p denotes the chosen accuracy threshold (note that, as everywhere else, it is assumed that the analytic approximation (2.4) holds for $x^{(h)}(t) = A^{(h)}(t) \cos \phi^{(h)}(t)$). Then it can be shown [7] that, given $\omega_f^{(1)}$, $\Delta\omega_f^{(1)}$, the frequency range for the hth harmonic will be

$$[\omega_f^{(h)} - \Delta\omega_f^{(h)}/2, \omega_f^{(h)} + \Delta\omega_f^{(h)}/2] : \omega_f^{(h)} = h\omega_f^{(1)}, \ \Delta\omega_f^{(1)} \lesssim \Delta\omega_f^{(h)} \lesssim h\Delta\omega_f^{(1)}. \tag{4.14}$$

Based on (4.14) and the scaling properties of the WFT/WT, one can now determine the region within which the optimal resolution parameter $f_0^{(h)}$ for the hth harmonic lies. Given that the first harmonic was extracted from the TFR calculated with resolution parameter $f_0^{(1)}$, and assuming that the corresponding time and frequency resolutions are appropriate for this first harmonic, one has

$$\textbf{WFT: } f_0^{(h)} \in [f_0^{(1)}/h, f_0^{(1)}], \quad \textbf{WT: } f_0^{(h)} \in [f_0^{(1)}, hf_0^{(1)}]. \tag{4.15}$$

In numerical implementation, the optimal $f_0^{(h)}$ is searched for by first breaking the region (4.15) into N_r values $f_{0;r=1,\dots,N_r}^{(h)}$ (the default is $N_r = 10$). For each of them one calculates the TFR in the corresponding frequency range (4.8), extracts the harmonic from it (see Sect. 4.1.2), estimates the corresponding consistency $\rho_r^{(h)}$ (4.4), and tests the harmonic for being true (see Sect. 4.1.3). Among the $f_{0;r}^{(h)}$ for which the harmonic was identified as true, the one characterized by the highest $\rho_r^{(h)}$ is selected. It is then used as the initial value for an iterative golden section search of the optimal $f_0^{(h)}$ (with default accuracy being $\epsilon_f = 0.01$), maximizing the consistency $\rho^{(h)}$ (4.4).

Note that (4.15) does not take into account the interference between harmonics and with the other components which might lie nearby in frequency, so that in some cases the upper bound on $f_0^{(h)}$ might be higher than (4.15); the same consideration applies to the lower bound. Therefore, if near the minimum or maximum of the tested $f_0^{(h)}$ the consistency $\rho^{(h)}$ is found to grow, the search is continued in that direction until the peak appears.

Remark 4.2.1 The same procedure is performed while checking whether the extracted component represents a first or a higher harmonic by extracting and testing its $h = 1/2, 1/3, \ldots$ harmonic candidates (see Sect. 4.1.4). The formulas (4.14) and (4.15) remain valid in this case, but the upper and lower bounds for $\Delta\omega_f^{(h)}$ in (4.14) and $f_0^{(h)}$ in (4.15) change places [7].

4.2.2 Improved Reconstruction of Nonlinear Modes

Given the reconstructed amplitudes $A^{(h)}(t)$, phases $\phi^{(h)}(t)$ and frequencies $\nu^{(h)}(t)$ of all the true harmonics, the most straightforward way to reconstruct the full NM is just to add all the $A^{(h)} \cos\phi^{(h)}(t)$ together. However, in this way the NM picks up noise contributions from all harmonics, which can make it quite inaccurate.

Fortunately, there is a cleverer way to perform the reconstruction, also yielding more accurate parameters for the individual harmonics. Thus, one can utilize the theoretical amplitude, phase and frequency relationships between the harmonics, i.e. $A^{(h)} = a_h A^{(1)}(t)$, $\phi^{(h)} - h\phi^{(1)} = \varphi_h$ and $\nu^{(h)}(t) = h\nu^{(1)}(t)$. Then, because the components with higher amplitudes are expected to be less noise-corrupted, one can refine the parameters of each harmonic by weighted averaging over the parameters of all harmonics:

$$\begin{aligned}
\tilde{A}^{(h)}(t) &= \langle A^{(h)}(t)\rangle \frac{\sum_{h'} A^{(h')}(t)}{\sum_{h'} \langle A^{(h')}(t)\rangle}, \\
\tilde{\phi}^{(h)}(t) &= \arg\Big[\sum_{h'} \min(1, h'/h)\langle A^{(h')}(t)\rangle e^{i(h\phi^{(h')}(t) - \overline{\Delta\phi_{h'h}} - 2\pi I[(\Delta\phi_{h'h}(t) - \overline{\Delta\phi_{h'h}})/2\pi])/h'}\Big], \\
\tilde{\nu}^{(h)}(t) &= \frac{\sum_{h'} \min(1, h'/h)\langle A^{(h')}(t)\rangle h\nu^{(h')}/h'}{\sum_{h'} \min(1, h'/h)\langle A^{(h')}(t)\rangle},
\end{aligned} \tag{4.16}$$

where $\Delta\phi_{h'h}(t) \equiv h\phi^{(h')}(t) - h'\phi^{(h)}(t)$ and $\overline{\Delta\phi_{h'h}} \equiv \arg\langle e^{i\Delta\phi_{h'h}(t)}\rangle \in [-\pi, \pi]$, while $I[\ldots]$ denotes rounding to the nearest integer, so that $I[0.8] = 1$, $I[-0.6] = -1$ (the corresponding term is needed to eliminate possible noise-induced phase slips, i.e. the rapid growth by 2π in the phase differences). Note also the multiplier $\min(1, h'/h)$, appearing for phase and frequency refinement in (4.16). It is needed to account for the scaling of phase and frequency errors of lower harmonics when they are mapped to higher ones. Thus, if $\nu^{(1)}(t)$ has an error $\varepsilon(t)$, then $h\nu^{(1)}(t)$ will have error $h\varepsilon(t)$.

Remark 4.2.2 The expressions in (4.16) are based on the assumption that the reconstruction error for each harmonic is directly proportional to its amplitude. However, this is true only if, first, there are no side components with which harmonics interfere and, secondly, the amount of noise picked up while reconstructing harmonics is the same for each of them. These criteria are rarely both satisfied in practice, so in general one should take into account the mean absolute reconstruction errors $c(h)$ of the hth harmonic parameters. This can be done by changing $A^{(h')}(t) \to c^{-1}(h')A^{(h')}(t)$ in all the expressions (4.16). Unfortunately, the errors $c(h)$ are very hard to estimate, even roughly, and therefore they are not used.

The refinement (4.16) not only makes the reconstructed NM much more accurate, solving the problem of picking up the cumulative noise of all the harmonics, but also reduces the noise level in each harmonic separately. Thus, noise-induced variations in the parameters of different harmonics are expected to be mutually independent, and so they average out on being added together. As a result, the more harmonics are contained in the NM, the more accurately it can be reconstructed. While NMD is generally noise-robust, due to being based on time-frequency methods for which only the spectral power of noise in the vicinity of components' frequencies matters, NM reconstruction by (4.16) raises this robustness to an extreme extent.

Remark 4.2.3 It should be noted that the refinement (4.16) is not valid if the NM waveform changes in time, i.e. the relationships between the amplitudes of the harmonics and their phase shifts are non-constant. In this case one should just sum all extracted harmonics, even though some of them might be highly corrupted. Nevertheless, in this work it is assumed that the waveform is constant, as otherwise even the harmonic identification scheme discussed in Sect. 4.1.3 might become inaccurate (though if the waveform changes slowly enough, it still works well).

4.3 Parameters and Their Choice

There are different settings that can be used while applying NMD. However, many of the parameters are either set at well-established values, or can be chosen adaptively, thus removing the ambiguity and making NMD a kind of superadaptive method. The parameters and their choices are summarized in Table 4.1. They are discussed in more detail below.

4.3.1 Resolution Parameter f_0

The resolution parameter f_0 of the window/wavelet determines the trade-off between the time and frequency resolution of the WFT/WT (see Sect. 2.22). For harmonics, the adaptive choice of f_0 has been already discussed in Sect. 4.2.1, but it is still unclear how to select the resolution parameter at the first step, i.e. when one extracts the main component (Sect. 4.1.1). This is, however, a fundamental issue related to the general use of the TFRs that is inherent to all methods which directly or indirectly rely on them.

In practice, one needs to select it based on a compromise between better reflecting time variability and better resolving components in frequency, so the optimal choice depends on the characteristics of the components contained in the signal. Unfortunately, at present there does not seem to be any universal approach enabling one to choose the resolution parameter appropriately for any given signal (see [6, 19] for a discussion of this issue and the effects of different choices); it therefore remains the

Table 4.1 The basic NMD parameters (given by names under which they are implemented in the codes [18]), their default choices and descriptions

Parameter	Default	Description
'TFRtype'	WT first, then chosen adaptively (see Sect. 4.3.2)	The type of TFR (WFT or WT) to use for extracting the modes
'ModeNum'	chosen adaptively (see Sect. 4.1.5)	The number of modes to extract
'SurrType' 'SurrNum' 'Signif'	'FT' 40 0.95	The type of surrogates, their number and the significance threshold used for testing the residual against noise (see Sect. 4.1.5)
'S'	3	The required number of consecutive false harmonics to stop harmonics extraction (see Sect. 4.1.4)
'S0'	3	Same as 'S', but for the fractional $(1/n)$ harmonics that are used to determine whether the extracted component is fundamental harmonic or not (see Sect. 4.1.4)
'WCons'	[1,1,0]	The weighting factors $[w_A, w_\phi, w_\nu]$ used for calculating the harmonics' overall consistencies (4.4), see Sect. 4.1.3
'SNumber' 'SLevel'	100 0.95	The number of surrogates and the significance threshold used for testing the harmonics for being true (see Sect. 4.1.3)
'CLevel'	determined by (4.7)	The minimum consistency required for harmonic to be true (see Remark 4.1.3)
'MinSupp'	0.001	The precision used for determining the minimal support (3.12), and therefore for estimating the component's frequency range in the TFR (see Remark (3.2.4))
'RecMethod'	chosen automatically (see Sect. 4.3.3)	Reconstruction method used for estimating the parameters of the harmonics
'AdaptRange' 'AdaptRes'	determined by (4.15) 10	The range in which to search for f_0 for harmonics and the initial number of guesses to check (see Sect. 4.2.1). Additional accuracy settings can also be specified

Note that this table does not include parameters of the WFT/WT itself, which are discussed in Chap. 2

only important parameter of NMD that cannot be adaptively selected. The choice of f_0, however, seems to be slightly more universal for the WT, because the latter adjusts its resolution in relation to the frequency band being studied; for the WT with lognormal wavelet (2.19) or Morlet wavelet (2.18), one usually uses $f_0 = 1$, setting some standard limits on the allowable relative amplitude/frequency modulation of the components and the frequency distances between them.

Remark 4.3.1 Note, that the very possibility of adjusting time and frequency resolution is a great advantage of the TFR-based methods, and NMD in particular. Thus, many other techniques, e.g. (ensemble) empirical mode decomposition [20, 21], do not allow this possibility, having time and frequency resolutions which are fixed around some implicit values [22, 23]. The choice of f_0 therefore gives NMD more flexibility in comparison to methods not possessing such a parameter.

4.3.2 TFR Type: WFT or WT?

The main difference between the WFT and the WT lies in the type of frequency resolution (linear for the former and logarithmic for the latter), and the preference for one type of TFR over the other depends on the signal structure (see Sect. 2.2.3 for a detailed discussion of this and related issues). Without some a priori knowledge, it is hard to select adaptively the most apporiate resolution type, especially given the associated problems related to the choice of the resolution parameter (see previous subsection). However, after some component has been extracted (even roughly), based on its properties one can judge whether it can be better represented by the WFT or by the WT. Thus, if the time-variability of the component's parameters increases with frequency, then the WT is the most suitable, whereas otherwise one should prefer the WFT. Given the initially extracted component's amplitude $A(t)$, phase $\phi(t)$ and frequency $\nu(t)$, an empirical condition for selecting the most appropriate TFR type can be stated as:

$$\begin{aligned} \Big(1 + V[\partial_t \nu(t), \nu(t)]\Big)^{-1} + \Big(1 + V[\partial_t A(t), \nu(t)]\Big)^{-1} < 1 \;\Rightarrow\; \text{use WFT;} \\ \text{otherwise} \;\Rightarrow\; \text{use WT,} \end{aligned} \tag{4.17}$$

where

$$V[x(t), y(t)] \equiv \frac{\operatorname{std}\big[|x(t)/y(t)|^{+}\big]}{\operatorname{std}\big[|x(t)/\langle y(t)\rangle|^{+}\big]}, \quad \Big(|f(t)|^{+} \equiv |f^{+}(t)| \Big). \tag{4.18}$$

Thus, the values of $V[\partial_t A(t), \nu(t)]$ and $V[\partial_t \nu(t), \nu(t)]$ quantify whether the amplitude and frequency modulation of the component become stronger with increasing frequency ($V < 1$) or not ($V > 1$). In the former case a reliable representation of the component requires higher time resolution for higher frequencies, so the

WT is preferred, while for the latter case the WFT should be used. For example, for linear ($\nu(t) = \nu_0 + at$) and hyperbolic ($\nu(t) = \exp(at)$) chirps one has $V[\partial_t \nu(t), \nu(t)] = \infty$ and $V[\partial_t \nu(t), \nu(t)] = 0$, reflecting the well-known [19] fact that the WFT and WT are most appropriate for their representation, respectively.

The derivatives $\partial_t \nu(t)$ and $\partial_t A(t)$ in (4.17) can be estimated by numerical differentiation. However, when noise is present in $A(t)$ and $\nu(t)$, it will be greatly amplified in such estimates, so they will be generally quite noisy. Consequently, instead of standard deviations std$[x(t)]$ in (4.18), it is better to use 75-percentiles, i.e. the width of the range within which 75 % of the values of $x(t)$ are contained. Alternatively, one can reconstruct $\partial_t \nu(t)$ and $\partial_t A(t)$ by deriving the direct reconstruction formulas for them as explained in Chap. 7.3.

The remained question is which TFR type to use for the preliminary signal exploration, i.e. extraction of the component's parameters to be used in (4.17). As was discussed, the answer depends on the signal structure, and there is no universal criterion. However, since the WT is usually faster to calculate (due to its logarithmic frequency scale) and generally has a more universal choice of the resolution parameter (see previous subsection), it is used by default as an initial guess.

Summarizing, one first calculates the WT of the signal and extracts from it the component and its parameters. Then the criterion (4.17) is utilized to determine which is the best type of TFR to use in the given case. If this is the WT, one retains the component already extracted; otherwise, one calculates the WFT of the signal in the corresponding frequency range (3.13) and re-extracts all the parameters from it. To preserve the time and frequency resolution properties for which the component was extracted from the WT (which are assumed to have been appropriate for it), the resolution parameter for the WFT $f_0^{(WFT)}$ should be adjusted accordingly. If the WFT and WT are calculated using a Gaussian window (2.12) and lognormal wavelet (2.19), the rule is [7]

$$f_0^{(WFT)} \approx 2\pi f_0^{(WT)} / \langle \nu(t) \rangle . \tag{4.19}$$

Since for all harmonics by definition one would have identical $V[\partial_t \nu(t), \nu(t)]$ and $V[\partial_t A(t), \nu(t)]$ in (4.17), the same type of frequency resolution (linear or logarithmic) is appropriate for all of them. Hence, for the extraction of harmonics one should use the same type of TFR as was selected for the original component.

Remark 4.3.2 The logarithmic frequency resolution of the WT can by itself introduce correlation between the frequency and the amplitude/frequency variations of the component. For example, when the signal is corrupted with white noise, its power in the WT will be proportional to frequency, leading to larger noise-induced amplitude/frequency variations in the extracted component at higher frequencies. Such artificial correlation might cause the criterion (4.17) to select the WT as the preferred representation even when this is not the case. To avoid this, the threshold 1.1 instead of 1 is used by default in (4.17).

4.3.3 Reconstruction Method: Direct or Ridge?

As discussed in Sect. 3.2, one can use either of two alternative methods for reconstruction of the components and their parameters from the WFT/WT: direct (3.8) or ridge (3.5). The ridge method is more noise-robust, but the direct method allows time variability in the component's parameters to be followed more accurately at low noise levels. The optimal parameter estimates for the fundamental component extracted at the first step (Sect. 4.1.1) can be chosen adaptively as described in Sect. 3.2.4. For harmonics, on the other hand, the appropriate reconstruction method can be determined simply as that giving the highest consistency $\rho^{(h)}$ (4.4). Therefore, while adapting the resolution parameter for the harmonic, at each f_0 its parameters are reconstructed by both methods and at the end the one characterized by highest $\rho^{(h)}$ is selected.

For the other parts of the NMD procedure it is inherently better to use a particular reconstruction method. Thus, in the criterion for selecting the TFR type (4.17) the ridge estimates of amplitude $A(t)$ and frequency $\nu(t)$ are preferred due to their noise-robustness. Additionally, while testing the signal against noise in the stopping criterion (Sect. 4.1.5), the discriminating statistics D (4.9) appears to be slightly more powerful if calculated using ridge estimates. This seems natural because the curve extraction is based on the amplitudes and frequencies of the peaks (see Sect. 3.1), though the noise-robustness of ridge reconstruction is advantageous here as well.

4.3.4 Other Parameters

All the other parameters can be partitioned into two groups: some pre-fixed settings and the parameters of numerical accuracy. The former group includes the significance levels for the two surrogate tests (used for identification of the harmonics, Sect. 4.1.3, and for the stopping criterion, Sect. 4.1.5), the minimal consistency $\rho_{\min}$ (4.7) etc. Each of these parameters is set to a well-established value, either corresponding to a standard convention (such as 95 % significance of the surrogate tests [2, 3]) or found empirically (e.g. the expression (4.7) for $\rho_{\min}$).

The second group includes the accuracy with which to determine the optimal f_0 while adapting it to the harmonics, the precision ϵ_s to use for determining the minimal support (3.12), etc. Here one faces the usual tradeoff between accuracy and computational cost. The default values, however, are sufficient in most cases and further increase of precision leads to only a slight improvement in the accuracy of the method.

4.4 Summary

In this Chapter NMD has been introduced and developed, with all its subprocedures and related issues being considered in detail. The method is intended to be adaptive, which is achieved by matching its parameters to the signal's structure (Sect. 4.3); and very robust to noise, as a result of the general noise-robustness of the TFR-based approaches, and also due to improvements introduced in Sect. 4.2. Furthermore, by identifying and joining together all harmonics corresponding to the same mode (see Sects. 4.1.2 and 4.1.3), NMD is able to recover the oscillations within any waveform. Finally, by stopping the decomposition when the signal obtained at the next iteration does not pass the surrogate test against noise (Sect. 4.1.5), the method retrieves only physically meaningful oscillations, with trends and noise being left as the residual.

The full NMD procedure can be summarized as follows:

1. The signal is detrended and tested against the null hypothesis of noise (Sect. 4.1.5). If it does not pass the test, the procedure is stopped.
2. The WT (Sect. 2.2.2) of the signal is calculated, and the most dominant component is extracted from it (Sect. 4.1.1). Based on the properties of this component, the optimal TFR type—WFT or WT—is selected (Sect. 4.3.2); it is then used in all the subsequent steps. The component is re-extracted from the signal's WFT if this type was determined to be the more appropriate than the WT.
3. The component obtained in the previous step generally represents a particular harmonic of some NM, but not always the first one. To find the latter, the component's $h = 1/2$ possible harmonic is extracted (Sect. 4.1.2) and tested for being true (Sect. 4.1.3); this is done using the WFTs/WTs calculated for different resolution parameters f_0, with the most appropriate among them being selected as described in Sect. 4.2.1. The same procedure is performed for $h = 1/3, 1/4, \ldots$ harmonic candidates, until three of them in a row are identified as false. The fundamental harmonic of the NM to which the component belongs is then taken as the true harmonic with the smallest h (or as the original component if all $h = 1/2, 1/3, 1/4$ harmonics were identified as false).
4. Based on the parameters of the fundamental harmonic, its possible $h = 2, 3, \ldots$ harmonics are extracted and tested for being true in qualitatively the same way as was done for $h = 1/2, 1/3, \ldots$ in the previous step.
5. Using the parameters of all true harmonics, the full NM is reconstructed as described in Sect. 4.2.2. It is then subtracted from the signal, and steps 1–5 are repeated for the residual.

The choice of the reconstruction method—direct (3.8) or ridge (3.5)—for each step and substep of the procedure are discussed in Sect. 4.3.3. The operation of NMD will also be illustrated on simulated examples in the next Chapter. The codes for running NMD are freely available [18].

References

1. H.-T. Wu, Instantaneous frequency and wave shape functions (i). Appl. Comput. Harmon. Anal. **35**(2), 181–189 (2013)
2. T. Schreiber, A. Schmitz, Surrogate time series. Physica D **142**(3), 346–382 (2000)
3. J. Theiler, S. Eubank, A. Longtin, B. Galdrikian, J.D. Farmer, Testing for nonlinearity in time series: the method of surrogate data. Physica D **58**(1), 77–94 (1992)
4. R.G. Andrzejak, A. Kraskov, H. Stögbauer, F. Mormann, T. Kreuz, Bivariate surrogate techniques: necessity, strengths, and caveats. Phys. Rev. E **68**(6), 066202 (2003)
5. R.Q. Quiroga, A. Kraskov, T. Kreuz, P. Grassberger, Performance of different synchronization measures in real data: a case study on electroencephalographic signals. Phys. Rev. E **65**(4), 041903 (2002)
6. D. Iatsenko, P.V.E. McClintock, A. Stefanovska, Linear and synchrosqueezed time-frequency representations revisited: overview, standards of use, reconstruction, resolution, concentration, and algorithms. Digital Signal Processing (in press) 2015. doi:10.1016/j.dsp.2015.03.004
7. D. Iatsenko, P.V.E. McClintock, A. Stefanovska, Nonlinear Mode decomposition: a new noise-robust, adaptive decomposition method [preprint—arXiv:1207.5567], 2014
8. J.D. Farmer, E. Ott, J.A. Yorke, The dimension of chaotic attractors. Physica D **7**(1), 153–180 (1983)
9. P. Grassberger, Generalized dimensions of strange attractors. Phys. Lett. A **97**(6), 227–230 (1983)
10. H. Hentschel, I. Procaccia, The infinite number of generalized dimensions of fractals and strange attractors. Physica D **8**(3), 435–444 (1983)
11. P. Grassberger, I. Procaccia, Characterization of strange attractors. Phys. Rev. Lett. **50**(5), 346–349 (1983)
12. P. Grassberger, I. Procaccia, Measuring the strangeness of strange attractors. Physica D **9**(1), 189–208 (1983)
13. J. Theiler, Efficient algorithm for estimating the correlation dimension from a set of discrete points. Phys. Rev. A **36**(9), 4456 (1987)
14. J. Theiler, D. Prichard, Constrained-realization Monte-Carlo method for hypothesis testing. Physica D **94**(4), 221–235 (1996)
15. C. Stam, J. Pijn, W. Pritchard, Reliable detection of nonlinearity in experimental time series with strong periodic components. Physica D **112**(3), 361–380 (1998)
16. J. Theiler, T. Lookman, Statistical error in a chord estimator of correlation dimension: the "rule of fiv". Intern. J. Bifurc. Chaos **3**(3), 765–771 (1993)
17. T. Schreiber, A. Schmitz, Improved surrogate data for nonlinearity tests. Phys. Rev. Lett. **77**(4), 635 (1996)
18. The NMD Toolbox for MatLab, which contains all the needed codes, is freely available at http://www.physics.lancs.ac.uk/research/nbmphysics/diats/nmd/ (see also http://www.physics.lancs.ac.uk/research/nbmphysics/diats/tfr/)
19. S. Mallat, *A Wavelet Tour of Signal Processing*, 3rd edn. (Academic Press, Burlington, 2008)
20. N.E. Huang, Z. Shen, S.R. Long, M.C. Wu, H.H. Shih, Q. Zheng, N. Yen, C.C. Tung, H.H. Liu, The empirical mode decomposition and the Hilbert spectrum for nonlinear and non-stationary time series analysis. Proc. R. Soc. Lond. A **454**, 903–995 (1998)
21. Z. Wu, N.E. Huang, Ensemble empirical mode decomposition: a noise-assisted data analysis method. Adv. Adapt. Data Anal. **1**(1), 1–41 (2009)
22. M. Feldman, Analytical basics of the EMD: two harmonics decomposition. Mech. Syst. Sig. Proc. **23**(7), 2059–2071 (2009)
23. G. Rilling, P. Flandrin, One or two frequencies? The empirical mode decomposition answers. IEEE Trans. Sig. Proc. **56**(1), 85–95 (2008)

Chapter 5
Examples, Applications and Related Issues

After describing all parts of the NMD procedure in the previous Chapter, the method is now applied to both simulated (Sect. 5.1) and real (Sect. 5.2) examples.

5.1 Simulated Examples

To assess reliably the performance of NMD and compare it to that of the other methods, this section considers specific simulated signals whose composition is precisely known, so that the extracted NMs can be compared with the true ones.

5.1.1 Example 1

The first, simple and illustrative, example is taken as the signal depicted in Fig. 5.1a. Its WFT and WT are shown in Fig. 5.1b, c, respectively. From the figure it is immediately evident that, while in the WFT the harmonics with frequencies around 3 and 4 Hz are distinguishable, in the WT they interfere strongly and cannot be resolved; in contrast, the WT has high enough time resolution at 7 Hz to represent the frequency modulation of the corresponding harmonic, while in the WFT this highest harmonic self-interferes (there appear "bridges" between consecutive frequency modulation cycles), indicating that the time resolution is insufficient. Therefore, for the present case one cannot appropriately extract all harmonics using either WFT or WT with constant resolution parameter f_0. However, adaptive representation of the harmonics, as discussed in Sect. 4.2.1, solves the problem.

Nonlinear mode decomposition proceeds as follows. First, the WT of the signal is calculated, and the dominant component is extracted from it as described in Sect. 4.1.1; in result one obtains the first harmonic of the NM (located around 1 Hz). The extracted component is then tested with the surrogates against noise (see Sect. 4.1.5); in the present case, it passes the test (100 % significance) and is therefore

D. Iatsenko, *Nonlinear Mode Decomposition*,
Springer Theses, DOI 10.1007/978-3-319-20016-3_5

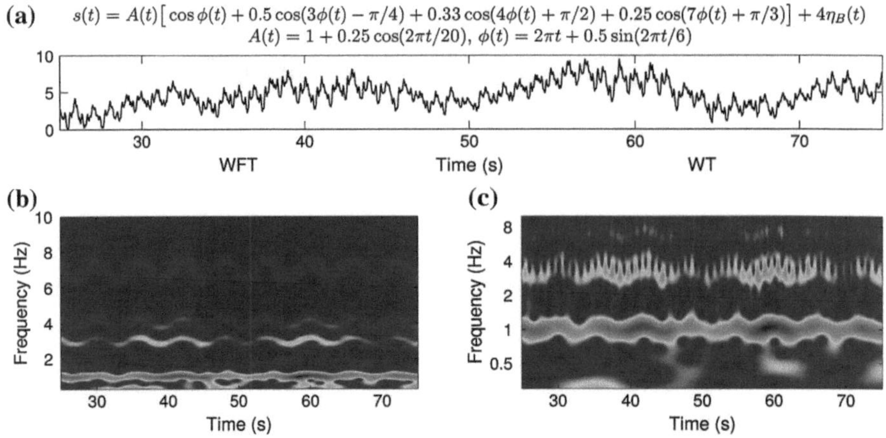

Fig. 5.1 **a** The central 50s section of the signal $s(t)$ specified by the equation at the *top*, which represents a single NM corrupted by Brownian noise of standard deviation equal to 4 (Brownian noise of unit deviation $\eta_B(t_n) \sim \sum_{m=1}^{n} \eta_W(t_m)$ is obtained as the normalized cumulative sum of the Gaussian white noise signal $\eta_W(t)$). **b**, **c** Central parts of the WFT and WT, respectively, of the signal $s(t)$ shown in (**a**). The signal was sampled at 100 Hz for 100 s

regarded as genuine. By application of the criteria (4.17), the WFT is determined to be more suitable for the representation of this component than the WT, and is used in what follows. Thus, the component is re-extracted from the corresponding signal's WFT (see Sect. 4.3.2), and its parameters are reconstructed by the both direct (3.8) and ridge (3.5) methods. Using (3.11), it is established that the ridge estimates of amplitude, phase and frequency seem to be the more accurate in the present case, and are therefore taken as the ones to be used.

Next, the extracted component is tested for being the first harmonic by checking its $1/n$ harmonic candidates. Thus, first the $1/2$ harmonic is extracted and tested for being true (see Sect. 4.1.3) using different resolution parameters f_0 within the range (4.15); among the direct and ridge estimates obtained for f_0 at which the harmonic was identified as true, the ones maximizing the consistency (4.4) are chosen. In the present case, the $1/2$ harmonic is identified as false for all tested f_0, so it is discarded. The same procedure is performed for 1/3 and 1/4 harmonics, which are both identified as false. Since there are $S_0 = 3$ consecutive false harmonics (1/2, 1/3 and 1/4), it is correctly concluded that the extracted component is first (and not higher) harmonic of the NM.

The higher harmonics $h = 2, 3, \ldots$ are then extracted and tested in qualitatively the same way as was done for $h = 1/2, 1/3, 1/4$. If some harmonic is identified as true, it is subtracted from the signal to remove its interference with the other harmonics. As a result of the procedure, all genuine harmonics $h = 3, 4, 7$ are correctly identified as true, and all the others as false. The resolution parameters are adapted for each harmonic so as to optimally represent and reconstruct it, as discussed

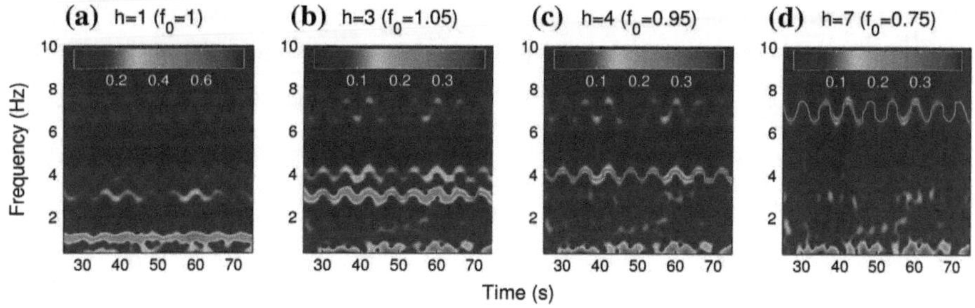

Fig. 5.2 The WFTs from which each true harmonic is reconstructed, with the corresponding extracted ridge *curves* being shown by *solid magenta lines*. The window resolution parameter f_0 is adjusted individually for each harmonic. After the harmonic is reconstructed, it is subtracted from the signal, so that e.g. the first harmonic no longer appears in the WFTs of (**b**–**d**). Note that the color scaling in (**b**–**d**) covers half the amplitude range of that in (**a**)

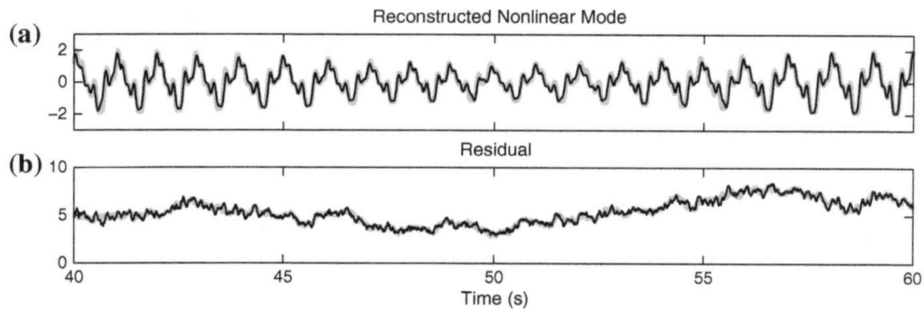

Fig. 5.3 The result of applying NMD to the signal shown in Fig. 5.1a: **a** the reconstructed NM (*black line*) is compared with the true NM (*gray background line*); **b** the residual provided by NMD (*black line*) is compared with the actual background noise (*gray background line*)

in Sect. 4.2.1 and illustrated in Fig. 5.2 for the present case. Harmonic extraction is stopped when $S = 3$ consecutive harmonics $h = 8, 9, 10$ are identified as false.

After reconstructing all harmonics, their parameters are refined using (4.16) and they are then summed up to form the full nonlinear mode. This NM is subtracted from the signal, and the procedure is repeated on the residual. However, the component extracted from it does not pass the surrogate test against noise (see Sect. 4.1.5), and therefore NMD is stopped, with one NM being extracted and the residual correctly identified as noise (in the present case Brownian). The result of NMD is shown in Fig. 5.3, from which one can see that even the residual Brownian noise is well-recovered. To the best of my knowledge, there is at present no other method that can decompose even the current relatively simple signal (shown in Fig. 5.1a) in such an accurate and physically meaningful way.

For example, the results of EMD [13] and EEMD [45] procedures are shown in Fig. 5.4. In contrast to NMD, (E)EMD produces 13 distinct components, with only the first harmonic of the NM being more-or-less reliably recovered in one of them.

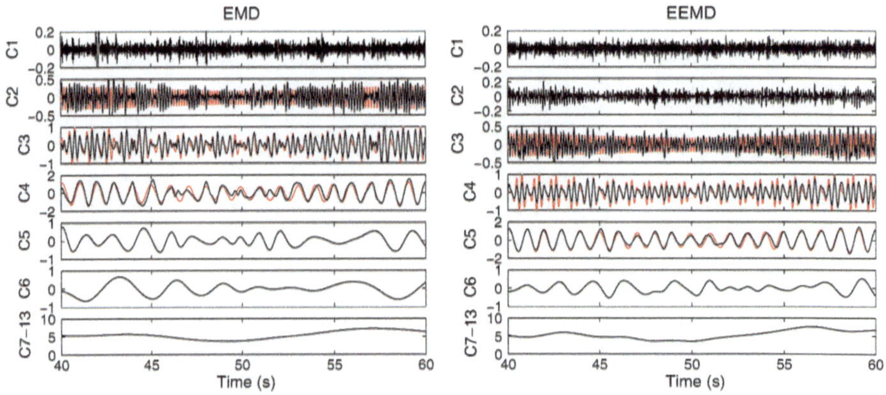

Fig. 5.4 The result of applying EMD (*left panels*) and EEMD (*right panels*) to the signal shown in Fig. 5.1. *Red lines*, where present, show the real 1st harmonic (in C4 for EMD and C5 for EEMD), sum of the 3rd and 4th harmonics (in C3 for EMD and C4 for EEMD) and the 7th harmonic (in C2 for EMD and C3 for EEMD). The *bottom panels* show the sum of the components 7–13. For EEMD, 1000 Gaussian white noise realizations were used, with their standard deviations being ten times smaller than that of the signal

Thus, in Fig. 5.4 C4 for EMD and C5 for EEMD represent the first harmonic, C3 for EMD and C4 for EEMD is the noise-spoiled mix of the 3 and 4 harmonics, C2 for EMD and C3 for EEMD is the badly corrupted 7th harmonic (with influence from harmonics 3 and 4 in the case of EEMD), while none of the other "components" has any physical meaning at all.

Remark 5.1.1 Interestingly, (E)EMD [13, 45] has logarithmic frequency resolution [10, 30], and therefore suffers from the same drawbacks as the WT in relation to the representation of harmonics. This is why it merges harmonics 3 and 4 into a single component (as these harmonics are merged in the WT as well, see Fig. 5.1). However, while for the WT its logarithmic frequency resolution can be tuned by changing f_0, for (E)EMD the resolution is fixed around some implicit value [10, 30].

5.1.2 Example 2

The exceptional noise-robustness of NMD and the power of the surrogate test in distinguishing true from false harmonics can be demonstrated by considering the signal shown together with its WFT and WT in Fig. 5.5. For this signal, the harmonics of the second NM are located exactly at the places where the harmonics of the first NM are expected to be, so that they can easily be confused with them. Moreover, it is very hard to distinguish true from false harmonics in the present case because each NM has constant amplitude and only very small frequency modulation (the absolute deviation between the expected frequency of the second harmonic of the

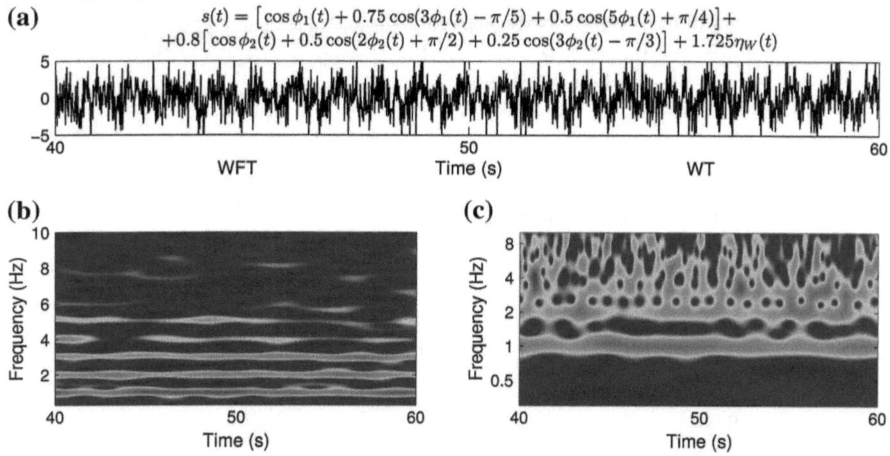

Fig. 5.5 **a** The central 20 s section of the signal $s(t)$ specified by the equation at the top, representing the sum of two NMs corrupted by Gaussian white noise of standard deviation equal to 1.725; the phases of the NMs $\phi_{1,2}(t)$ were obtained as $\phi_{1,2}(t) = \int_0^t \nu_{1,2}(\tau)d\tau$ with $\nu_{1,2}(t)/2\pi = (1, 2) + 0.01\tilde{\eta}_{B;1,2}(t)$, where $\tilde{\eta}_{B;1,2}(t)$ are two independent realizations of unit deviation Brownian noise filtered in the range [0.01, 0.2] Hz. **b**, **c** Central parts of the WFT and WT of the signal $s(t)$ shown in (**a**). The signal was sampled at 100 Hz for 100 s

first NM and the frequency of the first harmonic of the second NM is only $|2\nu_1^{(1)}(t) - \nu_2^{(1)}|/2\pi = 0.016 \pm 0.014$ Hz). Furthermore, the noise is exceedingly strong, in standard deviation being 1.5 times that of the full noise-free signal, 1.8 times that of the first NM, 2.7 times that of the second NM, and 12.2 times that of the smallest (3rd) harmonic of the second NM, located at around 6 Hz (the latter is buried under the noise and not even visible in the WFT shown in Fig. 5.5b).

Because the noise is white in the present case, its power in the WT grows with frequency, as seen from Fig. 5.5 (note, however, that such a situation rarely arises for real signals). Consequently, instead of using the WT and then adaptively selecting the appropriate type of TFR, for the present case it is better to use the WFT from the very beginning. Nonlinear mode decomposition then proceeds as usual: it first extracts the dominant component and tests it against noise; if it passes the test, NMD extracts its harmonics and identifies the true ones; then the full NM is reconstructed and subtracted from the signal, after which the procedure is repeated on the residual until it does not pass the surrogate test against noise.

The relevant information about the NMD outcome is summarized in Table 5.1. As can be seen, NMD correctly identifies all harmonics and joins them into two distinct modes. Moreover, the amplitude ratios a_h and phase shifts φ_h for each NM (see (4.1)), calculated from the reconstructed amplitude and phases as $a_h = \langle A^{(h)}\rangle/\langle A^{(1)}\rangle$ and $\varphi_h \equiv \arg\left[\langle e^{i(\phi^{(h)} - h\phi^{(1)})}\rangle\right]$, are very close to their actual values; this is true even for the (buried in noise) 3rd harmonic of the second NM. The ridge method is automatically

Table 5.1 Summary of the results of NMD applied to the signal shown in Fig. 5.5

Harmonic number h	$\rho^{(h)}$	Signif. level	Rec. method	$f_0^{(h)}$	Amp. ratio a_h		Phase shift φ_h/π	
					True	Extr	True	Extr
First NM (significance against noise = 100 %)								
1	–	–	*Ridge*	*1.00*	*1*	*1*	*0*	*0*
2	0.41	50 %	Ridge	2.85	–	–	–	–
3	*0.75*	*100 %*	*Ridge*	*2.08*	*0.75*	*0.74*	*–0.20*	*–0.20*
4	0.29	69 %	Ridge	1.77	–	–	–	–
5	*0.63*	*98 %*	*Ridge*	*2.01*	*0.50*	*0.51*	*0.25*	*0.26*
6	0.11	52 %	Ridge	1.77	–	–	–	–
7	0.04	67 %	Direct	1.21	–	–	–	–
8	0.06	67 %	Ridge	0.41	–	–	–	–
Second NM (significance against noise = 100 %)								
1	–	–	*Ridge*	*1.00*	*1*	*1*	*0*	*0*
2	*0.68*	*98 %*	*Ridge*	*2.85*	*0.50*	*0.45*	*0.50*	*0.51*
3	*0.40*	*95 %*	*Ridge*	*1.71*	*0.25*	*0.29*	*–0.33*	*–0.33*
4	0.09	79 %	Ridge	1.95	–	–	–	–
5	0.11	91 %	Ridge	0.64	–	–	–	–
6	0.07	17 %	Ridge	1.00	–	–	–	–
Residual (significance against noise = 37 %)								

For each NM the significance of the surrogate test against noise (see Sect. 4.1.5) was based on 100 surrogates. The columns left-to-right provide information for the hth harmonic: the value of the consistency $\rho^{(h)}(1, 1, 0)$ (4.4); the significance of the test against independence (see Sect. 4.1.3); the method chosen for reconstruction of the harmonic (see Sect. 4.3.3); the resolution parameter $f_0^{(h)}$ adapted for the harmonic considered (see Sect. 4.2.1); the true amplitude ratio $a_h \equiv A^{(h)}(t)/A^{(1)}(t)$; the extracted amplitude ratio; the true phase shift $\varphi_h \equiv \phi^{(h)}(t) - h\phi^{(1)}(t)$; the extracted phase shift. A harmonic was identified as true if both $\rho^{(h)} \geq \rho_{\min} = 0.25$ (4.7) and the significance level is $\geq$95 %, with the corresponding rows being represented with italic. The numbers of real harmonics are emboldened, and it can be seen that only they were identified as being true. For simplicity, the results for $h = 1/n$ harmonics (which are tested to check whether the extracted component is the first harmonic) are not shown; in each case all three consecutive $h = 1/2, 1/3, 1/4$ harmonics were identified as false

selected for reconstructing of all harmonics, which is indeed the preferable choice due to its noise-robustness (see Sect. 3.2).

From Table 1 it can be noted that, for all true harmonics, the resolution parameter f_0 used is higher than the original. This is because the higher the f_0, the easier it is to segregate the component from noise [16]. However, increasing f_0 at the same time worsens the accuracy of representation of amplitude/frequency modulation of the component [16], so its choice is determined by a compromise between reflecting well the time-variability of component's parameters and suppressing the noise; the adaptive scheme that is used by NMD (see Sect. 4.2.1) effectively implements this criterion.

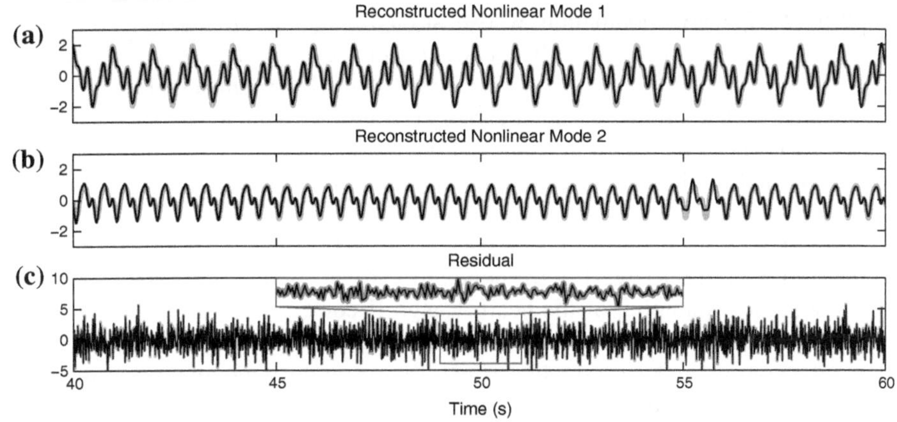

Fig. 5.6 The result of applying NMD to the signal shown in Fig. 5.5. In **a** and **b** *black lines* indicate the two reconstructed NMs and the *gray background lines* show the true NMs for comparison. **c** Similarly, the *black* and *gray lines* show the residual returned by NMD and true background noise, respectively

The final results of NMD are shown in Fig. 5.6. Both NMs are reconstructed with great accuracy, which is like a miracle given such strong noise corruption of the original signal (see Fig. 5.5); even the residual noise is recovered almost exactly. Such performance is unachievable with the other existing methods, e.g. (E)EMD in the present case produces 13 components, and none of them reliably represent any harmonic (not shown). Note, that NMD can produce even more accurate results if the resolution parameter is adjusted from the very beginning, i.e. for the first harmonics (and not only the higher ones). However, as discussed in Sect. 4.3.1, at present there does not seem to be a good and universal way of doing this.

5.2 Real Applications

After demonstrating its success in the analysis of simulated signals, NMD and the related techniques are now applied to real data.

5.2.1 Decomposing Human Blood Flow Signals

The decomposition of skin blood flow signals, measured non-invasively by laser-Doppler flowmetry (LDF) [28], is a very tough task, with no method at present being able to do it well. This was demonstrated in [12] for the example of Karhunen-Loève decomposition, while (E)EMD usually also fails. Thus, blood flow signals contain a large amount of "physiological" noise, as well as having components with amplitude/frequency modulation whose strength and speed change with time. Nonetheless, as will be seen, NMD can often tackle even such complicated cases.

Blood flow signals (here and in what follows, "blood flow" means skin blood flow recorded by LDF) exhibit oscillatory activity at multiple frequencies, and the WT has been found especially useful in studies of their structure [38]. Six oscillations have been identified in blood flow and attributed to different physiological mechanisms [21, 34, 38, 39], with characteristic frequency ranges of (approximately): 0.6–2 Hz (I), 0.15–0.6 Hz (II), 0.04–0.15 Hz (III), 0.02–0.04 Hz (IV), 0.01–0.02 Hz (V) and 0.005–0.01 Hz (VI). Range I corresponds to cardiac oscillations, which originate from rhythmical pumping of the heart. Range II corresponds to respiratory oscillations, which are the consequence of the mechanical influence of respiration on the cardiac output and, to a smaller extent, the respiratory modulation of the heart rate [17, 32]. The mechanism underlying range III oscillations, which are present in most hemodynamic signals, is not generally agreed: in studies of cardiac activity and blood pressure variability they are usually attributed to the sympathetic nerve activity, being regarded as a result of time-delays in the baroreflex feedback loop [18, 26], while the many authors studying microvascular blood flow relate these oscillations to myogenic activity of the smooth muscle cells [34, 38, 39]. Finally, the oscillations in the IV, V and VI ranges were attributed to the neurogenic, NO-dependent endothelial and NO-independent endothelial activity [21, 34, 38, 39], respectively.

Armed with this knowledge, one can now apply NMD. To better utilize the prior information, the procedure is applied to each of the above-mentioned physiological frequency ranges individually, starting from the first one. Thus, for a given range the dominant component is first extracted and tested against noise; if it passes the test, one then extracts its harmonics, reconstructs the NM and subtracts it from the signal; the procedure is repeated for the same range until the next extracted component does not pass the test against noise.

The results of the procedure are shown in Figs. 5.7 and 5.8 for the examples of two blood flows. Clearly, NMD is able to decompose these signals into physically meaningful oscillations with complex waveforms (and it also returns their amplitudes, phases and frequencies).

In both cases, it was possible to extract the cardiac component (around 1 Hz), while activity in ranges IV, V and VI did not pass the test against noise. However, in the example of Fig. 5.7 there are strong sympathetic/myogenic oscillations (around 0.1 Hz), which were extracted, while there is no apparent activity in the frequency range II (respiratory) (Fig. 5.7b). In contrast, for the example of Fig. 5.7 the respiratory oscillations are present and the sympathetic/myogenic are buried under noise. The waveforms of the cardiac oscillations are also different in two cases.

We have found that, in practice, NMD is almost always able to extract the cardiac component accurately from the blood flow signal using the default settings though, to improve the accuracy and speed of the method, it is advisable to filter the signal in the corresponding frequency range before applying NMD. The respiratory component, on the other hand, can be extracted only in a limited number of cases on account of often being very small in amplitude compared to the physiological noise inherent in blood flow signals. Thus, one can always obtain "some" component from the

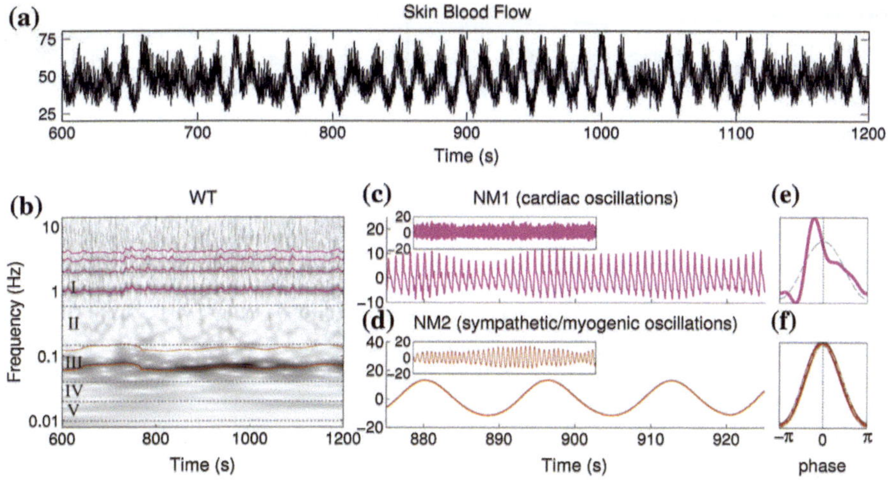

Fig. 5.7 **a** An example of a blood flow signal, measured by LDF with the probe over the right wrist *caput ulna* (for more details, see [34, 39]); the signal was sampled at 40 Hz for 1800s, and the panel shows its central 600 s part. **b** The WT of the signal shown in (**a**); for better illustration, *black*-and-*white* colorcode is used in contrast to all other figures. *Gray-dotted lines* partition the frequency axis into the regions within which the physiologically meaningful oscillations are located (according to [21, 34, 38, 39], see text). *Bold colored lines* indicate those extracted components which pass the surrogate test against noise, with *thinner lines* of the same color showing their higher harmonics. **c**, **d** The reconstructed NMs, with the main graph showing them during 50s, and small insets—during 600 s (as in (**a**)); the colors of the lines correspond to those of the *curves* in (**b**). **e**, **f** The waveforms of the oscillations shown in (**c**) and (**d**), respectively; the *gray dashed line* shows a pure sinusoidal waveform of amplitude equal to that of the first harmonic, and is provided as a guide to the eye. The cardiac waveform **e** has four harmonics $h = 1, 2, 3, 4$ with $a_h = [1, 0.52, 0.37, 0.16]$ and $\varphi_h/\pi = [0, 0.4, 0.66, -0.97]$ (in the notation of (4.1)); the waveform in **f** is characterized by $h = 1, 2$, $a_h = [1, 0.08]$ and $\varphi_h/\pi = [0, 0.07]$

corresponding frequency range, but it is often very inaccurate and consequently fails the surrogate test. Nevertheless, if the respiratory signal is also available, then one can use it as a reference signal for extracting the corresponding oscillations accurately from the blood flow signal even where they are very weak. An approach for doing this is discussed in Sect. 5.2.3 below; it can also be used to improve the extraction of cardiac oscillations if the corresponding ECG, from which cardiac phase can be extracted more accurately than from the blood flow signal, is available. Finally, although we were able to extract sympathetic/myogenic oscillations (range III) in some cases, they are typically very hard to extract, as are also those at even lower frequencies (ranges IV–VI).

In general, the properties and presence of the oscillations in blood flow varies from subject to subject, being influenced by many factors, such as the state of the microvasculature (which might be influenced by age and gender), properties of the skin etc. Clearly, NMD can be very useful for the study and classification of these effects, and one example of its possible clinical use will be considered in the next subsection.

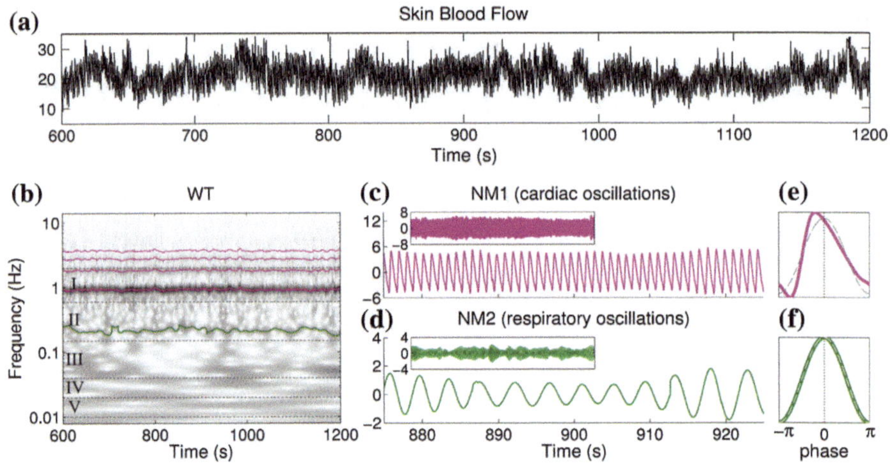

Fig. 5.8 Same as in Fig. 5.7, but for the blood flow measured from a different subject. The cardiac waveform (**e**) has four harmonics $h = 1, 2, 3, 4$ with $a_h = [1, 0.32, 0.15, 0.04]$ and $\varphi_h/\pi = [0, 0.41, 0.95, -0.44]$ (in the notation of (4.1)); the respiratory oscillations have only one harmonic, so that the waveform in (**f**) is a pure sinusoid

Remark 5.2.1 As discussed in Sect. 4.1.5, even if the component extracted from a particular frequency range does not pass the surrogate test against noise (as e.g. for ranges IV, V and VI in the above examples), this does not necessarily mean that there is no physiologically meaningful activity there. Thus, the underlying oscillations might be simply very small so that they are easily masked by noise, which is often the case for the respiratory oscillations. The other possibility is that the resolution parameter used is not appropriate to represent reliably the component of interest. This is often the case for sympathetic/myogenic oscillations, which might change their amplitude and/or frequency very rapidly at certain times. In fact, the best choice in such situations would probably be some time-varying $f_0(t)$, but its form is generally very hard to choose.

5.2.2 Clinical Application: Cardiac Waveform Study

In clinical practice, different health aspects are examined through the analysis of various signals generated by the human body. For example, the state of the cardiovascular system is usually assessed by analysing signals such as the electrocardiogram (ECG), blood pressure, blood flow (measured on a different sites of the body) etc. The properties of these signals have both prognostic and diagnostic value for many diseases [25, 27, 44], including cardiac failure and hypertension, which are among the today's most important health problems. In particular, essential hypertension—a chronic elevation of blood pressure of unclear origin that represents a major risk

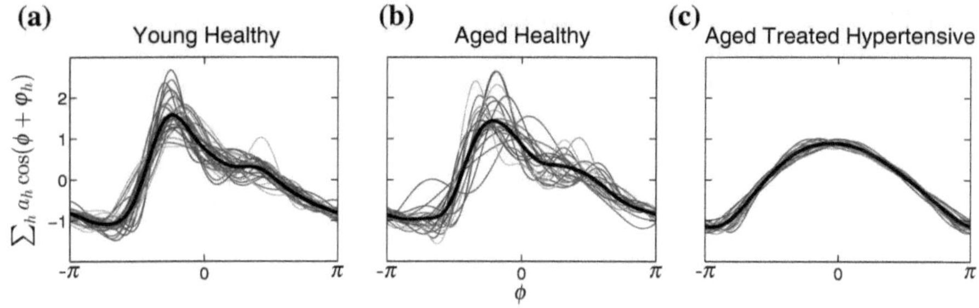

Fig. 5.9 The cardiac waveform (as recovered by NMD from the skin blood flow) for the three groups: **a** young healthy; **b** aged healthy; **c** aged treated hypertensive. *Thin colored lines* show waveforms for individual subjects, while *thick black lines* correspond to group averages

Table 5.2 Summary of subject characteristics for each group (Y—young healthy, A—aged healthy, ATH—aged treated hypertensive)

	Y	$p_{\text{Y-A}}$	A	$p_{\text{A-ATH}}$	ATH
Gender (M/F)	15/14	0.45	9/13	0.38	12/10
Age (years)	24.4 ± 3.4	0.00	71.1 ± 6.6	0.93	70.3 ± 6.7
SBP (mmHg)	118.2 ± 16.2	0.36	123.7 ± 12.5	0.02	138.8 ± 16.4
BMI (kg/m^2)	22.7 ± 2.7	0.17	23.7 ± 2.5	0.00	27.2 ± 3.9

The values of p indicate the significance of the Wilcoxon rank sum test for each parameter. SBP—systolic blood pressure, BMI—body mass index

factor for heart attack, stroke and many other complications—affects more than 30 % of the UK population, being one of the most widespread health issues.

As was demonstrated in the previous subsection, NMD can be very useful for studying blood flow signals. In particular, using NMD one can accurately recover the corresponding cardiac oscillations, their variability and waveform. This allows the investigation of various related aspects, such as the effects of ageing, hypertension and antihypertensive treatment on the properties of cardiac oscillations in the microvasculature.

As an example, Fig. 5.9 presents the cardiac waveforms found by NMD for the three groups: 29 young healthy subjects (group Y); 22 aged healthy subjects (group A); and 22 aged treated hypertensives (group ATH). The parameters for each group are shown in Table 5.2, while more detailed information about the third group (ATH) is presented in Table 5.3; an unpaired two-sided Wilcoxon rank sum test is used for comparisons between the groups, with statistical significance being assumed for $p < 0.05$.

From Fig. 5.9 it is clear that the form of the cardiac pulses in skin blood flow, while changing only slightly with age, is affected considerably by treated hypertension. This effect is statistically significant, as follows from the analysis of the relationships between cardiac harmonics for the three groups, presented in Fig. 5.10. Hence, although current antihypertensive treatment indeed reduces blood pressure and regresses many hypertension-associated changes [9, 22, 35, 43], it apparently

Table 5.3 *Left two columns* clinical characteristics of the aged treated hypertensive (ATH) group, apart from those listed in Table 5.2

Parameter	Mean ± SD	Medication	No. of subjects
Time since diagnosis (years)	10.0 ± 6.2	Beta-blockers	10
Total cholesterol (mmol/l)	4.26 ± 1.22	ACE inhibitors	10
HDL cholesterol (mmol/l)	1.36 ± 0.38	ARB	4
LDL cholesterol (mmol/l)	2.32 ± 0.99	CCB	9
Triglycerides (mmol/l)	1.31 ± 0.53	Diuretics	8
hs-CRP (mg/l)	2.62 ± 2.03	Statins	16
Capillary refill time (s)	2.5 ± 0.6	Aspirin	12
Ankle brachial pressure index	1.08 ± 0.09		
Height (m)	1.68 ± 0.10		
Weight (kg)	77.6 ± 16.5		

Right two columns the number of subjects within the ATH group treated with each type of medication; the majority were taking multiple medications. ARB—angiotensin receptor blockers, CCB—calcium channel blockers

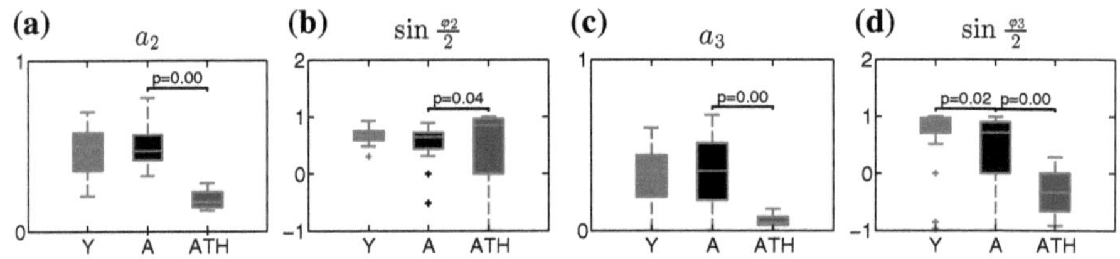

Fig. 5.10 Distributions of the amplitude ratios a_h and the sines of halved phase shifts $\sin\frac{\varphi_h}{2}$ (in the notation of (4.1)) of cardiac harmonics $h = 2$ (**a**, **b**) and $h = 3$ (**c**, **d**) for the three groups: Y—young healthy, A—aged healthy, ATH—aged treated hypertensive. The value of p indicates significance of the Wilcoxon rank sum test for the corresponding distributions; it is shown only if the difference is significant ($p < 0.05$). The comparison is made only between groups Y and A, and between A and ATH

does not restore everything back to normal. The physiological interpretation of the above results, however, is a separate topic lying out of the scope of the present work.

Remark 5.2.2 Note that the cardiac waveform is only one out of many blood flow properties to which one can gain access with NMD. For example, the amplitude/frequency variabilities of different flow oscillations can also be studied. Additionally, given the reconstructed cardiac mode, one can accurately estimate the pulse transit times (time differences between the R-peaks in the electrocardiogram and the next cardiac peaks in the flow), which represent another clinically relevant characteristic.

5.2.3 Removing Cardiac Artifacts from a Human EEG Signal

Nonlinear mode decomposition can alternatively be used to filter the signal from an extraneous oscillatory activity, provided that there is an associated signal from which the phase and frequency of the latter can accurately be extracted. Thus, using the NMD harmonic extraction procedure, the fundamental harmonic of the extraneous mode can be extracted as the $h = 1$ harmonic of the reference component based on its (reference) phase and frequency. In this way the resolution parameter $f_0^{(1)}$ is adjusted from the start, allowing one to represent this fundamental oscillation well even if it is strongly corrupted by noise or other influences. This application of NMD will be illustrated on the example of removing cardiac artifacts from the human electroencephalogram (EEG) recording, using the electrocardiogram (ECG) as the reference signal.

The EEG often contains artifacts related to heart activity, which arise due to blood flow pulsations underneath the probes (the so-called ballistocardiogram (BCG) artifacts [36]) and possibly also due to direct pick up of heart electrical activity. The BCG artifacts are extremely prominent in the EEG measured in a magnetic field (e.g. simultaneously with a magnetic resonance imaging scan), in which case they are usually filtered out by independent component analysis (ICA) [5, 14]. However, ICA requires as many simultaneous EEG measurements as possible, with cardiac artifacts being prominent and taking a similar form in most of them, which is not always the case in practice. It will obviously fail to remove the artifacts if only one EEG and one ECG signal are available, as the form taken by cardiac activity in the EEG is completely different from its form in the ECG signal (see below), contradicting the assumption of linear mixing on which ICA is based. In fact, we have found that ICA fails even given four EEGs, which is probably because the artifacts in them are relatively small, being hard to distinguish (although still capable of affecting some time-frequency measures); and, additionally, because the form and magnitude of these artifacts might be different in different EEGs (perhaps dependent on probe position). The (E)EMD method, too, fails to provide meaningful results in the present case.

Simultaneously measured EEG and ECG time series for the same subject are presented in Fig. 5.11a, b, with their WFTs for the default resolution parameter $f_0 = 1$ being shown in (c) and (d), respectively. Clearly, for $f_0 = 1$ the cardiac harmonics are not well distinguishable in the EEG's WFT (Fig. 5.11c), so that the extracted curve might not be very accurate. However, the cardiac phase $\phi^{(1)}(t)$ and frequency $\nu^{(1)}(t)$ can of course be extracted *directly* from the WFT of the ECG (Fig. 5.11d). They should be the same for the first cardiac harmonic in the EEG and in ECG, because both activities obviously have the same rhythmicity; however, the corresponding amplitudes might be different, and perhaps not even proportional to each other. This is because, depending on the measurement and the environment, the same activity might undergo various transformations that can change its amplitude dynamics and the corresponding waveform, but leave phase dynamics largely unaltered. For example, nonlinear modes $c(t) = A(t)[\cos\phi(t) + a\cos 2\phi(t)]$ and $\exp[c(t)]$ will have the

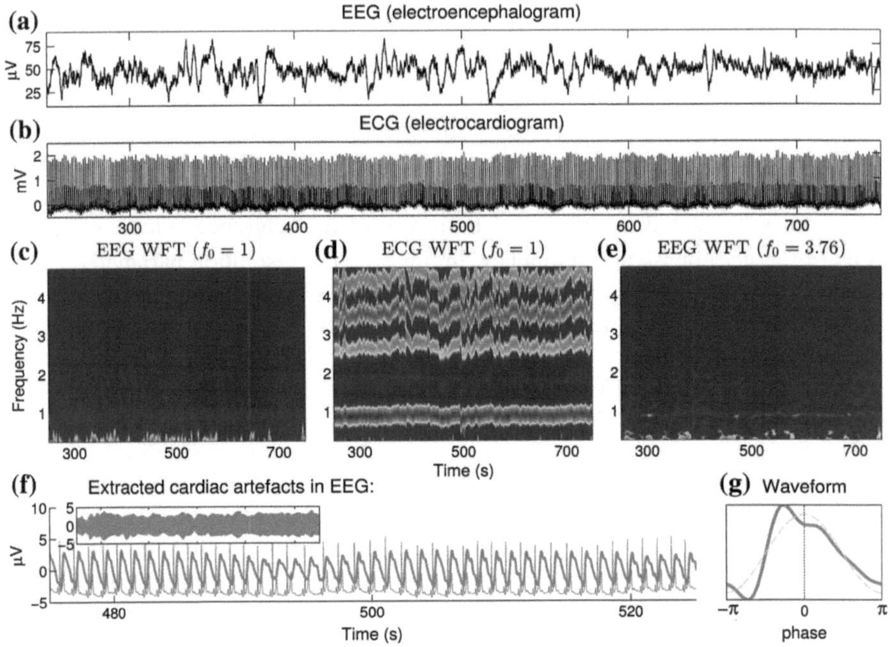

Fig. 5.11 **a** The EEG signal (measured using a BIS electrode placed on the forehead, as described in [40]). **b** The ECG signal (3-lead, with electrodes on shoulders and the lowest left rib, see e.g. [24]). **c**, **d** The WFTs of the EEG and ECG signals shown in (**a**) and (**b**), respectively, calculated using the default resolution parameter $f_0 = 1$. **e** The WFT of the EEG signal calculated using the adaptively adjusted resolution parameter $f_0 = 3.76$, due to which the cardiac component becomes much more visible than in (**c**). **f** The cardiac artifacts extracted from the EEG, with an inset showing them over the same 500 s as presented in (**a**, **b**); *gray background lines* show the ECG scaled to the dimensions of the plot. **g** The waveform of the cardiac artifacts, which has four harmonics with $a_h = [1, 0.33, 0.24, 0.11]$ and $\varphi_h/\pi = [0, 0.49, 1, -0.59]$ (in the notation of (4.1)); the *gray-dashed line* represents a pure sinusoid with the amplitude of the first harmonic, for comparison. The full signals, of duration 20 min. each, were recorded simultaneously and were sampled at 80 Hz; the EEG was actually measured under anaesthesia, but the same artifacts arise under all conditions

same fundamental phase and frequency, but different amplitude dynamics (i.e. the ratio between the amplitudes of the corresponding fundamental components will be time-varying) and different relationships between the harmonics.

Therefore, as mentioned above, the main cardiac component in the EEG can be extracted as the $h = 1$ harmonic of the reference cardiac component found from the ECG. Using the phase $\phi^{(1)}(t)$ and frequency $\nu^{(1)}(t)$ of the latter, this can be done in the usual manner (see Sect. 4.1.2), i.e. selecting the peaks nearest to the expected frequency profile $\nu^{(1)}(t)$ in the WFT/WT of EEG. The extracted first harmonic is then tested for being true (Sect. 4.1.3), and the procedure is repeated using different resolution parameters $f_0^{(1)}$ to find the most appropriate one (Sect. 4.2.1). However, because of the (above mentioned) possible discrepancy between the amplitude dynamics of the related components in different signals, the consistency measure

$\rho^{(h)}(w_A, w_\phi, w_\nu)$ (4.4) should be taken not as the default amplitude-phase consistency $\rho^{(1)}(1, 1, 0)$, but as a simple phase consistency $\rho^{(1)}(0, 1, 0)$; note that the threshold (4.7) becomes $\rho_{\min}(0, 1, 0) = 0.5$ in this case. Finally, if the extracted first harmonic is regarded as true, one finds its higher harmonics and reconstructs the corresponding NM in the usual way, i.e. using the unmodified procedures with $\rho^{(1)} = \rho^{(1)}(1, 1, 0)$ (see Sects. 4.1.2, 4.1.3 and 4.2).

Remark 5.2.3 Note that, in general, there might be a time delay between related oscillations in different signals, e.g. ECG and cardiac artifacts in EEG. However, unless it exceeds the characteristic time of the amplitude/frequency variations or the minimal surrogate time shift (see Sect. 4.1.3), it should not influence the results significantly. Nevertheless, one can adjust the timings of both components by maximizing the frequency consistency $\rho^{(1)}(0, 0, 1)$ (4.4) between them.

The EEG's WFT adapted (by maximizing phase consistency $\rho^{(1)}(0, 1, 0)$) for representation of the cardiac component is shown in Fig. 5.11e. Clearly, the corresponding ridge curves have become much more visible than in the default WFT presented in Fig. 5.11c. The cardiac artifacts extracted from the EEG are shown in Fig. 5.11f. Their waveform, presented in Fig. 5.11g, very much resembles the shape of the cardiac waves in the blood flow (cf. Fig. 5.8), but not that of the ECG. This is an indication of the BCG mechanism by which these cardiac artifacts are generated, which is also supported by the fact that their strength (and even their shape) might be different in EEGs from different probes for the same subject. Note that, depending on the particular EEG measurement, the artifacts might be inverted.

There are many other possible applications of NMD-based filtering. Thus, as mentioned previously, one can use this approach to extract the cardiac and respiratory components from the blood flow signal more accurately by utilizing reference phases obtained from the corresponding ECG and respiration signals. NMD can also be used to check whether a given signal contains oscillatory activity related to another one: if there is no oscillation originating from the same source as the reference signal, then the corresponding first harmonic extracted from a given signal based on the reference phase and frequency will be regarded as false. For example, we have not found any respiratory-related activity in the EEG, implying that the measurement process is almost unaffected by breathing.

Since the situation when one signal contains components related to other signals is ubiquitous in real life, the NMD-based filtering is expected to be very useful in many contexts. A great advantage of this approach is that it does not require the related oscillations in different signals to be of the same form (as is the case of e.g. ICA), but only to have the same phase dynamics.

5.2.4 Other Applications

In general, NMD can serve as an initial preprocessing which needs to be performed prior to applying any of the numerous existing methods for studying monocomponent signals or their sets (for a review see [4]). Thus, having first decomposed the original signal into its constituent NMs, one can then investigate the latter using one or some of the huge diversity of available techniques. For example, the structure of the interactions between different modes can be recovered by applying Bayesian inference to the extracted phases [7, 37], as was done in [15] to reconstruct the cardiorespiratory interaction; in this application, the high accuracy of the phases returned by NMD is especially advantageous. Another important problem where the use of NMD as a preprocessing tool can be very relevant is the classification of the oscillations contained in the signal, which might yield valuable insights into the nature and properties of the underlying phenomena. This application will be discussed in more detail in Sect. 5.3.

5.3 Physical Interpretation and System Classification

By measuring signals generated by a real system, one can get access to a certain subset of its properties, but many aspects of system behavior usually remain unavailable and effectively hidden. For example, the ECG reflects the overall electrical activity of the heart, but not the other numerous features of its operation. The recorded signals can therefore be viewed as certain projections of system dynamics on the measurement apparatus. This raises a question regarding the physical interpretation of the modes retrieved by NMD which, together with the possibility of classifying the extracted oscillations, is discussed in the present section.

In what follows, it will be assumed that the underlying system which produced the signal represents a kind of dynamical system [1, 8, 19], i.e. its behavior is fully determined by a particular set of ordinary differential equations

$$\frac{d\mathbf{x}}{dt} = \mathbf{F}(\mathbf{x}) \tag{5.1}$$

where $\mathbf{x} = [x_1, x_2, ..., x_D]$ correspond to system coordinates, the vector field $\mathbf{F}(\mathbf{x}) = [F_1(\mathbf{x}), F_2(\mathbf{x}), ..., F_D(\mathbf{x})]$ is some continuous $\mathbb{R}^D \to \mathbb{R}^D$ function, and D is the dimensionality of the system. The resultant signal can then be represented as $s(t) = g(\mathbf{x}(t))$, where $g(\cdot)$ is the "measurement" function. The behavior of systems of the form (5.1) and the underlying theory were thoroughly discussed in [1, 8, 19], while the applications of the numerous methods for studying different kinds of such systems are reviewed in [2, 4].

For what follows, it is also of paramount importance for the following to clarify the notion of nonautonomicity. The system is called nonautonomous if its associated vector field contains explicit time-dependence ($\mathbf{F}(\mathbf{x}) \to \mathbf{F}(\mathbf{x}, t)$ in (5.1)); otherwise it is called autonomous. Note, however, that any nonautonomous system can be represented as a subsystem of some higher-dimensional autonomous structure. For example, consider the (nonautonomous) system

$$\begin{cases} \dfrac{dx_1}{dt} = x_2, \\ \dfrac{dx_2}{dt} = -x_1[(\omega_0 + b\omega_b \cos \omega_b t)^2 - b\omega_b^2 \sin \omega_b t], \end{cases} \tag{5.2}$$

which generates the signal $x_1(t) = A\cos(\omega_0 t + b \sin \omega_b t + \varphi)$, where A and φ are determined by the initial conditions $x_{1,2}(0)$. Choosing two additional variables $x_{3,4}$ to account for the terms $\sim \sin \omega_b t$, $\sim \cos \omega_b t$ in (5.2), one can rewrite the corresponding equations as

$$\begin{cases} \dfrac{dx_1}{dt} = x_2, \\ \dfrac{dx_2}{dt} = -x_1[(\omega_0 + \omega_b x_3)^2 + \omega_b x_4], \\ \dfrac{dx_3}{dt} = x_4, \quad x_3(0) = b, \\ \dfrac{dx_4}{dt} = -\omega_b^2 x_3, \quad x_4(0) = 0. \end{cases} \tag{5.3}$$

Thus, the nonautonomous system (5.2) can be viewed as part of the higher-dimensional autonomous system (5.3) starting from a particular set of initial conditions.

However, in real cases the parameters of the full (autonomous) system which produced the signals can rarely be estimated, so one usually has to analyse its nonautonomous parts. For example, one can measure different aspects of the heart activity, but this activity is obviously influenced by many other body processes (e.g. respiration), which are in turn influenced by the environment, making treatment of the heart as autonomous system practically impossible. For an additional illustration, consider (5.1) with the vector field $\mathbf{F}(\mathbf{x})$ perturbed by an (in theory infinite-dimensional) multivariate Gaussian white noise $\eta(t)$: for such a case the full autonomous system cannot be recovered in any finite dimensions, so that one needs to take explicit account of the time-dependent noise (in this limiting case of nonautonomicity the system is called stochastic). Generally, whether to treat the system as autonomous or nonautonomous depends on what approach is more suitable for its analysis in the context considered, which is largely determined by the amount of information available.

5.3.1 "Decomposability" of Different Systems

It is not fully clear how compatible are the dynamical (5.1) and NMD (4.2) views of the signal, and many questions regarding this issue remain to be addressed, namely: How can the results of NMD be interpreted physically within dynamical systems theory? Are they always interpretable in this sense? Is there any pattern in terms of the "decomposability" of different systems? To answer these questions and establish the link between the two views, in this section NMD is applied to systems of different kinds and the results are considered in detail. Performance studies of the numerous other methods in a similar settings can be found in [2, 4].

In general, there are three types of behavior that dynamical systems (5.1) can exhibit in the limit $t \to \infty$ [1, 8, 19]: (i) decay to a fixed point ($\mathbf{x}(t) = \mathbf{x}_0$); (ii) move periodically ($\exists T_p : \mathbf{x}(t+T_p) = \mathbf{x}(t)$) or quasi-periodically (the trajectory $\mathbf{x}(t)$ is a superposition of periodic movements over different dimensions—limit cycles); (iii) move chaotically (i.e. deterministically but in practice unpredictably, with the trajectory being very sensitive to the choice of initial conditions). The first type of behavior is trivial and is therefore of no interest for the present analysis, and only the latter two will be considered.

To understand the effects of different behaviors and system (non)autonomicity on the NMD performance, consider the forced Duffing oscillator [6, 20]:

$$\frac{d^2x}{dt^2}+\delta\frac{dx}{dt}+\alpha x+\beta x^3 = A\cos\omega t \Leftrightarrow \begin{cases} \frac{dx}{dt} = y, \\ \frac{dy}{dt} = A\cos\omega t - \delta y - \alpha x - \beta x^3. \end{cases} \tag{5.4}$$

The system (5.4) will be called autonomous for $A = 0$ and nonautonomous (in two dimensions) otherwise; it was also found to exhibit chaotic behavior in certain parameter regions, where it will be called chaotic. Figure 5.12 presents examples of NMD-based decomposition of $x(t)$ generated by the system (5.4) in different regimes.

Remark 5.3.1 To provide more information, the signals in Fig. 5.12, as well as in Figs. 5.13 and 5.14 below, are decomposed into 20 modes (whether meaningful or not); the decomposition is stopped, however, if the standard deviation of the residual is lower than 0.05 of that of the signal. Additionally, since one can now have noise-induced modes (i.e. those not passing the surrogate test against noise, see Sect. 4.1.5), the ridge curve is extracted using the functional (3.3) with $p_2 = 0$ instead of $p_2 = 1$ (other parameters remain the same). Finally, in order to make the decomposition faster, the fundamental component and all its harmonics are extracted from the WT and are reconstructed by the direct method (3.8) only (instead of choosing the TFR type and reconstruction method adaptively, see Sect. 4.3).

As illustrated in Fig. 5.12a, when the system is two-dimensional and autonomous the most complex behavior it can exhibit is some strictly periodic limit cycle (which might nevertheless be of complicated shape as a result of nonlinearities in the system); this corresponds to a single mode of constant amplitude and frequency (Fig. 5.12d, g, j). Adding nonautonomicity to the original system, one effectively increases its dimensionality, which can make its behavior more complex (Fig. 5.12b). For example, the characteristics of the original limit cycle might be changed, or quasi-periodic behavior can emerge (the latter resulting in multiple modes and/or time-variability of the modes' amplitudes and frequencies).

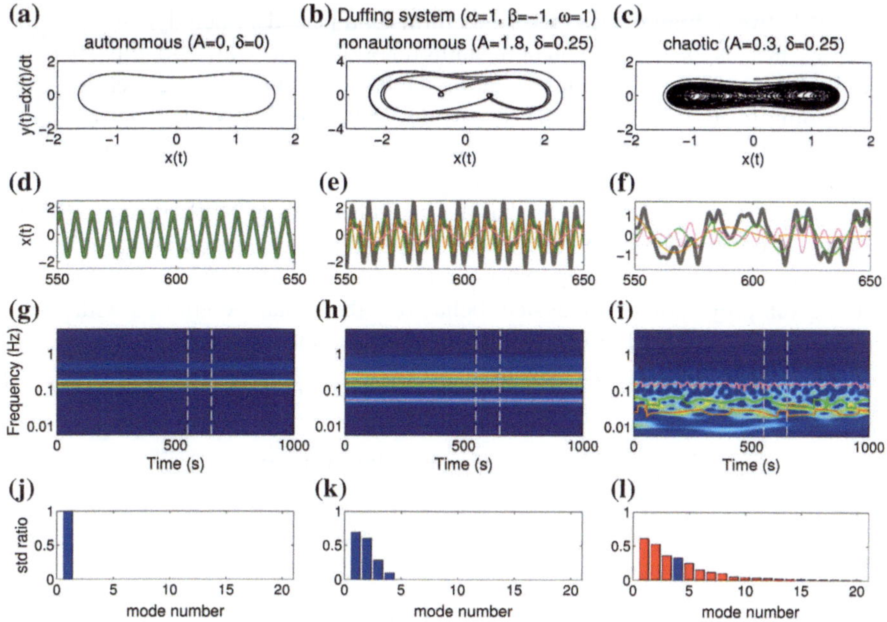

Fig. 5.12 The results of decomposition of the x-component of the Duffing system (5.4) in different regimes; parameters of the system for each regime are indicated on top of the figure. **a**, **b**, **c** The dynamics of the system (5.4) in the (x, y)-plane. **d**, **e**, **f** The original signal $x(t)$ (*thick background line*) and the three most prominent modes returned by NMD (*thin colored lines*), whether meaningful or not. **g**, **h**, **i** The wavelet transforms of the corresponding signals in (**d**, **e**, **f**); the *ridge curves* associated with fundamental harmonics of the three most prominent modes are indicated by *colored lines* (for simplicity, higher harmonics are not shown); *gray vertical dashed lines* indicate the time intervals for which the signals are presented in (**d**, **e**, **f**). **j**, **k**, **l** The ratios of standard deviations of each of the extracted NMs to the standard deviation of the original signal; *blue bars* indicate the modes which were regarded as meaningful (i.e. passed the surrogate test against noise, see Sect. 4.1.5), while *red bars* correspond to all the other ones. Starting from the initial conditions $x(0) = 0$, $y(0) = 1$, the system (5.4) was integrated using a Heun scheme with time-step of 0.01 s for 1000 s, and the resulting time-series were then resampled to 10 Hz. The adjustments to NMD procedure used for decomposing the signals presented in (**d**, **e**, **f**) are discussed in Remark 5.3.1

Remark 5.3.2 It should be noted, that the dynamics depicted in Fig. 5.12b, e can actually be represented as a single, fully periodic, mode with a very complex waveform. The reason why it is decomposed into multiple NMs (Fig. 5.12e, h, k) is that the frequency of this single mode is absolutely constant, which makes it nearly impossible to correctly identify all true harmonics (especially taking into account that some of them interfere in the TFR, see Fig. 5.12h). This is, however, a rather uncommon case.

On the other hand, TFR-based techniques do not seem to be suitable for analysing Duffing system in the chaotic regime (Fig. 5.12c, f, i, l). Thus, in this case the behavior of the system in the (x, y)-plane (Fig. 5.12c) can be perceived as rapid and frequent jumping between limit cycles of various forms and periods. Such dynamics cannot be represented as a superposition of meaningful oscillations with smoothly changing amplitudes and frequencies, thus violating the basic assumption (4.2) behind NMD and leading to its failure (Fig. 5.12f, i, l). Overall, Fig. 5.12 shows that the TFR-based methods usually work well for autonomous/nonautonomous systems, but are not always suitable for studying chaotic dynamics.

It should be stressed, however, that the inapplicability of NMD to some chaotic systems cannot be generalized to all cases. Thus, NMD tries to represent the system dynamics as a set of nonautonomous limit cycles, so everything depends on how well a given projection of the system behavior—the signal—conforms with such a representation. Figure 5.13 illustrates this important point on examples of the Lorenz [23] and Rossler [31] chaotic systems:

$$\textbf{Lorenz system:}\quad \begin{cases} \frac{dx}{dt} = \sigma(y - x), \\ \frac{dy}{dt} = x(\rho - z) - y, \\ \frac{dz}{dt} = xy - \beta z, \end{cases} \qquad \textbf{Rossler system:}\quad \begin{cases} \frac{dx}{dt} = -\omega y - z, \\ \frac{dy}{dt} = \omega x + ay, \\ \frac{dz}{dt} = b + z(x - c). \end{cases} \tag{5.5}$$

As can be seen, the projection of the Lorenz system dynamics on the x-axis (Fig. 5.13d) does not contain well-defined oscillations. This is because $x(t)$ captures all the exotic dynamical features, namely switching between the "wings" of the Lorenz attractor (see Fig. 5.13a, b) and the corresponding changes in the rotation direction (the latter being reflected by waveforms of the high-frequency oscillations, which are different for $x(t) < 0$ and $x(t) > 0$, see Fig. 5.13d). Due to this it is hard for NMD to decompose such a signal (Fig. 5.13d, g, j). In contrast, the z-coordinate of the Lorenz system (Fig. 5.13b, e) does not reflect the aforementioned aspects of the dynamics, which allows its approximate representation as a superposition of nonautonomous limit cycles and therefore leads to a better decomposition (Fig. 5.13e, h, k).

The signal from a chaotic system, however, does not necessarily need to miss some aspects of the dynamics to be well decomposable (i.e. admit the possibility of reliable mapping to the set of nonautonomous limit cycles). Thus, any component of the Rossler system (Fig. 5.13c) captures all features of its behavior but still can easily be decomposed by NMD, as illustrated in Fig. 5.13f, i, l for the x-component. Therefore, although TFR-based approaches are indeed often inapplicable for analysing chaotic dynamics, this is not a general rule.

Finally, one should be aware that the emergence of chaos is not the only complication that might spoil the results. Another (and probably more common) problem is noise which, being added to the right hand side of system equations (5.1), can significantly alter the unperturbed behavior. This is illustrated in Fig. 5.14 (cf. Fig. 5.12) for the example of the stochastic Duffing system:

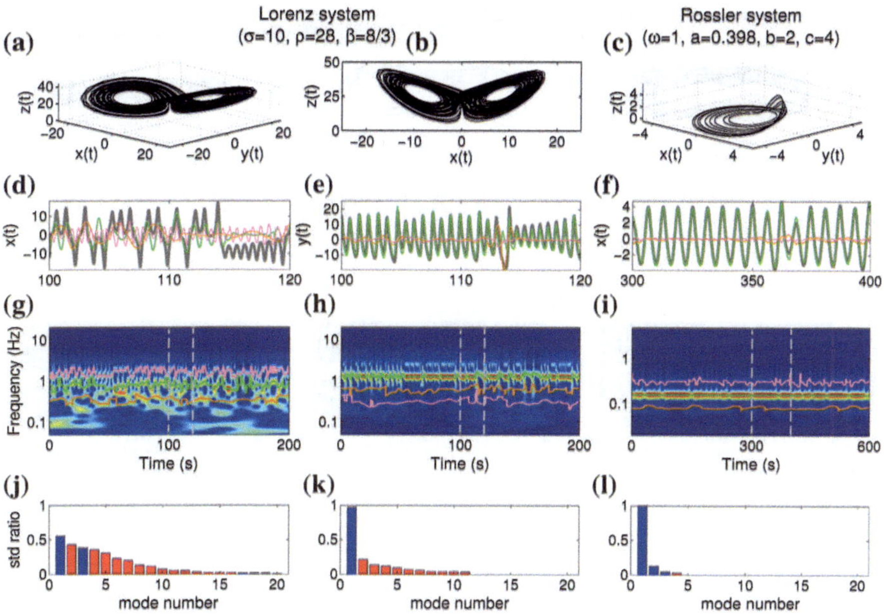

Fig. 5.13 The results of decomposition of the x- and z-component of the Lorenz system and z-component of the Rossler system (5.5); parameters of both systems are indicated on top of the figure. **a**, **c** The dynamics of the Lorenz and Rossler systems (5.5) in the full (x, y, z) space. **b** Projection of the Lorenz dynamics in (**a**) onto the (x, z)-plane. **d**, **e**, **f** The original signals to be decomposed (*thick background line*) and the three most prominent modes returned by NMD (*thin colored lines*), whether meaningful or not. **g**, **h**, **i** The wavelet transforms of the corresponding signals in (**d**, **e**, **f**); the *ridge curves* associated with fundamental harmonics of the three most prominent modes are indicated by *colored lines* (for simplicity, higher harmonics are not shown); *gray vertical dashed lines* indicate the time intervals for which the signals are presented in (**d**, **e**, **f**). **j**, **k**, **l** The ratios of standard deviations of each of the extracted NMs to the standard deviation of the original signal; *blue bars* indicate the modes which were regarded as meaningful (i.e. passed the surrogate test against noise, see Sect. 4.1.5), while red bars correspond to all the other ones. Lorenz and Rossler systems (5.5) were integrated using Heun scheme with time step of 0.01 for 200 and 600 s, respectively; the resulting time-series were then resampled to 40 Hz and to 10 Hz. The adjustments to NMD procedure used for decomposing the signals presented in (**d**, **e**, **f**) are discussed in Remark 5.3.1

$$\frac{d^2x}{dt^2}+\delta\frac{dx}{dt}+\alpha x+\beta x^3 = A\cos\omega t+\sigma\eta_W(t) \Leftrightarrow \begin{cases} \frac{dx}{dt} = y, \\ \frac{dy}{dt} = A\cos\omega t - \delta y - \alpha x - \beta x^3 + \sigma\eta_W(t), \end{cases} \tag{5.6}$$

where $\eta_W(t)$, as always, denotes white Gaussian noise of unit variance.

For the usual autonomous regime (Fig. 5.12a), adding dynamical noise changes the amplitude and frequency of the associated oscillations by causing transitions between similarly shaped limit cycles of different lengths (Fig. 5.14a). These noise-induced amplitude and frequency variations are, however, easily traced by NMD, and the signal can be decomposed reliably (Fig. 5.14d, g, j). In contrast, for the nonautonomous case (Fig. 5.12b) noise causes transitions between the limit cycles of very different

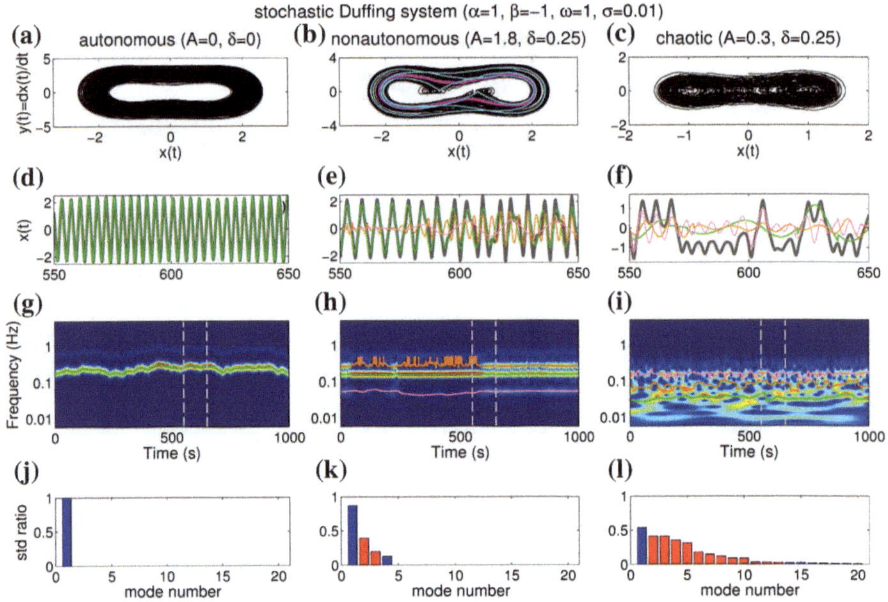

Fig. 5.14 The results of decomposition of the x-component of the stochastic Duffing system (5.6) in different regimes; parameters of the system for each regime are indicated on *top* of the figure. **a**, **b**, **c** The dynamics of the system (5.6) in the (x, y)-plane; *magenta* and *cyan curves* in (**b**) show approximate shapes of the limit cycles between which the dynamics switches. **d**, **e**, **f** The original signals to be decomposed (*thick background line*) and the three most prominent modes returned by NMD (*thin colored lines*), whether meaningful or not. **g**, **h**, **i** The wavelet transforms of the corresponding signals in (**d**, **e**, **f**); the *ridge curves* associated with fundamental harmonics of the three most prominent modes are indicated by *colored lines* (for simplicity, higher harmonics are not shown); *gray vertical dashed lines* indicate the time intervals for which the signals are presented in (**d**, **e**, **f**). **j**, **k**, **l** The ratios of standard deviations of each of the extracted NMs to the standard deviation of the original signal; blue bars indicate the modes which were regarded as meaningful (i.e. passed the surrogate test against noise, see Sect. 4.1.5), while *red bars* correspond to all the other ones. Starting from the initial conditions $x(0) = 0$, $y(0) = 1$, the system (5.6) was integrated using stochastic Heun scheme [11] with time-step of 0.01 s for 1000 s, and the resulting time-series were then resampled to 10 Hz. The adjustments to NMD procedure used for decomposing the signals presented in (**d**, **e**, **f**) are discussed in Remark 5.3.1

shapes (shown by magenta and cyan lines in Fig. 5.14b), thus making the dynamics inconsistent at different times (Fig. 5.14e). As a result, particular harmonics persist only during some time intervals and disappear during the others (Fig. 5.14h), making their identification and classification problematic and therefore causing problems for the method (Fig. 5.14k). Finally, in the chaotic regime (Fig. 5.12c) the behavior is already very complicated for time-frequency methods, so adding dynamical noise does not in fact change much in terms of decomposition (cf. Figs. 5.14f, i, l and 5.12f, i, l).

Summarizing, the success of decomposition of the signal generated by a system of the form (5.1), or its nonautonomous/stochastic version, depends on the type of behavior the system exhibits and the projection of this behavior onto the available data. Thus, NMD assumes that the signal can be represented as a superposition of physically meaningful oscillations (4.2) with amplitudes and frequencies that vary smoothly enough to satisfy the analytic approximation (2.4). As discussed previously, this set of assumptions does not include and is not included into, but rather intersects with, the set of assumptions of dynamical systems theory, which states that the system behavior can be described by a number of coupled differential equations (5.1).

Approaches based on either of these two views of the signal can produce spurious results if the corresponding assumptions are not satisfied, so that the choice of the right tool for analysing a given signal depends on its properties. Thus, it was demonstrated that NMD might fail in the case of chaotic behavior, which therefore tends to be harder to analyse using TFR-based methods as compared to other types of dynamics. However, no general statement can be made regarding this issue, as there are many other factors also influencing the results.

The methods of dynamical systems theory, on the other hand, can be inapplicable when the original system contains time-delays or other features not included in the description (5.1). Additionally, in practice the coordinates of the system are usually unavailable, so that one needs to reconstruct the underlying dynamics from the given set of measurements, which is not always an easy task [1, 8, 19]. This is especially true for nonautonomous and stochastic systems [2, 4], since e.g. the latter are in theory infinite-dimensional. For example, for the stochastic Duffing system in the autonomous regime (Fig. 5.14a), the noise-induced variations of the oscillations' amplitude and frequency are hard to estimate using the methods of dynamical systems theory. At the same time, they can accurately be recovered using NMD (Fig. 5.14d, g, j), which does not require reconstruction of the full coordinate space or any prior knowledge about the system.

5.3.2 Chronotaxic Systems and Their Identification

Recently, a new subclass of nonautonomous systems—*chronotaxic systems*—has been introduced and studied [3, 41, 42]. Roughly speaking (see [41, 42] for a more detailed definition), a system is chronotaxic if it is: (a) oscillatory (i.e. characterized by a limit cycle); and (b) its phase $\phi(t)$ does not just move freely along the cycle, as conventionally assumed, but is attracted to some $\phi_A(t)$, conferring the ability to resist external perturbations. Note, that this is exactly what one often observes in living systems, which are able to maintain their activity within physiological ranges even when strongly perturbed.

Chronotaxic behavior by definition satisfies the set of NMD assumptions (discussed in Sect. 5.3.1). Moreover, since the associated limit cycle is stable to perturbations, such adverse effects as jumping from one trajectory to the other (observed e.g. for the stochastic Duffing system in the nonautonomous regime, see Fig. 5.14b, e, h) cannot occur. Therefore, chronotaxic dynamics is always easily decomposable and, more generally, the TFR-based approaches are always appropriate for its analysis.

It is clearly desirable to be able to determine whether the originating systems generating different oscillations in the signal are chronotaxic or not. However, to classify these oscillations, one needs to study them separately, i.e. the signal should first be decomposed into its NMs. This can conveniently and accurately be achieved with NMD, and related techniques were used extensively in the chronotaxicity studies [3, 41]. Having found the phase $\phi(t)$ and frequency $\nu(t)$ of a particular mode, it can be tested for being chronotaxic with an approach suggested by Clemson et al. [3]. Basically, by applying different kinds of filters to the extracted phase one estimates the difference between the perturbed and unperturbed phases and then uses detrended fluctuation analysis (DFA) [29, 34] to examine the associated correlations, which are expected to differ between chronotaxic and non-chronotaxic systems.

The procedure can be summarized as follows:

1. Smooth the mode frequency $\nu(t)$ with a moving average of length τ_m seconds to obtain $\widetilde{\nu}(t)$:

$$\widetilde{\nu}(t) = \int \nu(\tau) m_{\tau_m}(\tau - t) d\tau, \tag{5.7}$$

   where $m_\tau(x \in [-\tau/2, \tau/2]) = 1$, $m_\tau(x \notin [-\tau/2, \tau/2]) = 0$. Note that, in practice, all signals are discrete, so that the integrals appearing in this and all other steps of the procedure should be discretized as well.
2. Construct the smoothed phase $\widetilde{\phi}(t)$ by integrating $\widetilde{\nu}(t)$:

$$\widetilde{\phi}(t) = \int_0^t \widetilde{\nu}(\tau) d\tau \tag{5.8}$$

3. Calculate an estimate of the difference $\delta\phi(t)$ between the perturbed and unperturbed phases as $\widetilde{\phi}(t) - \phi(t)$ detrended with a window of length τ_d seconds:

$$\delta\phi(t) = \widetilde{\phi}(t) - \phi(t) - \int [\widetilde{\phi}(\tau) - \phi(\tau)] m_{\tau_d}(\tau - t) d\tau \tag{5.9}$$

   The detrending is performed to eliminate "artificial" phase drifts that might be introduced by the previous two steps.
4. Perform DFA [29, 34] of $\delta\phi(t)$ and find the associated self-similarity parameter β. To do so, one constructs an integrated signal $Y(t) = \int_0^t \delta\phi(\tau)d\tau$, breaks it into non-overlapping segments each of length τ_s seconds, and subtracts the linear fits of every segment from $Y(t)$ (at the corresponding times) to form $\widetilde{Y}_{\tau_s}(t)$. The fluctuation strength $F(\tau_s)$ for the scale τ_s is then determined as

$$F(\tau_s) = \sqrt{\langle \widetilde{Y}^2_{\tau_s}(t) \rangle} \tag{5.10}$$

Finally, the DFA exponent β is estimated by fitting $\log F(\tau_s) = \alpha + \beta \log \tau_s$ in some range $\tau_s \in [\tau_s^{(1)}, \tau_s^{(2)}]$ (by default, 100 equilogspaced values of τ_s are used). Note that, to remove the boundary effects of the moving averages used in steps 1 and 3, it is better to perform DFA of only the central part $\delta\phi(\tau_m/2 + \tau_d/2 < t < T - \tau_m/2 - \tau_d/2)$, where T is the total length of the signal.

Parameter β returned by the method reflects the extent to which $\delta\phi(t)$ is correlated in time [29, 34]. For chronotaxic systems, the phase is attracted to some time-dependent $\phi_A(t)$, and therefore the perturbations to $\phi'(t)$ are expected to decay rapidly, implying short-range correlations in $\delta\phi(t)$ (small β); for non-chronotaxic systems, on the other hand, these perturbations will accumulate, leading to long-range correlations (high β) [3]. Within the assumption that perturbations to $\phi'(t)$ are represented by white Gaussian noise, for non-chronotaxic systems $\delta\phi(t)$ will take form of a Brownian walk ($\beta = 1.5$). It was therefore proposed [3] to regard the system as chronotaxic if $\beta \lesssim 1$.

Obviously, $\delta\phi(t)$ (5.9) represents a very rough approximation to the true difference between the perturbed and unperturbed phases. However, it still picks up well the correlation structure of the perturbations [3], allowing for a reliable analysis with DFA. The parameters $\tau_m, \tau_d, \tau_s^{(1,2)}$ of the method are generally chosen by the user, subject to a constraint $\tau_s^{(2)} \leq \min(\tau_m, \tau_d)$ (see [3] for detailed guidelines). However, the number of non-adaptive parameters can be reduced to two by setting $\tau_m = \tau_d = \tau_s^{(2)}$.

Figure 5.15 illustrates the use of NMD and the approach of [3] (outlined above) for system classification for the examples of chronotaxic oscillation $s_c(t) = \cos\phi_c(t)$ and a non-chronotaxic one $s_{nc}(t) = \cos\phi_{nc}(t)$; the corresponding phases $\phi_{c,nc}(t)$ are given by

$$\begin{aligned} \phi_c'(t) =& 1.25\omega_0(t)\sin(\phi_c(t) - \phi_0(t)) + 0.5\eta_W(t),\ \phi_0(t) = \int_0^t \omega_0(\tau)d\tau, \\ \phi_{nc}'(t) =& \omega_0(t) + 0.5\eta_W(t), \end{aligned} \tag{5.11}$$

where $\omega_0(t)/2\pi = 2 - (1/3)\cos(0.02t) + (1/3)\cos(0.01\pi t)$, $\phi_{c,nc}(0) = 0$, and $\eta_W(t)$ denotes white Gaussian noise of unit variance. Nonlinear mode decomposition is first utilized to find the oscillatory modes present in the signals together with the corresponding phases $\phi(t)$ and frequencies $\nu(t)$ (Fig. 5.15a, b); if used for the classification approach of [3], these phases and frequencies should be reconstructed by the direct method (Sect. 3.2.2), as it better picks up noise and therefore allows for a more reliable study of its correlation structure. One then calculates $\delta\phi(t)$ (5.9) (Fig. 5.15c, d), and uses DFA to estimate the associated self-similarity parameter β (Fig. 5.15e, f). As expected, for the chronotaxic system β is smaller than 1 (Fig. 5.15e), while for the non-chronotaxic system it is close to 1.5 (Fig. 5.15f).

The same scheme can be applied to real-life signals, e.g. to the NMs of both of the examples in Figs. 5.7 and 5.8. In these examples, no clear evidence of chronotaxicity was found for any of the corresponding oscillations. However, since the method

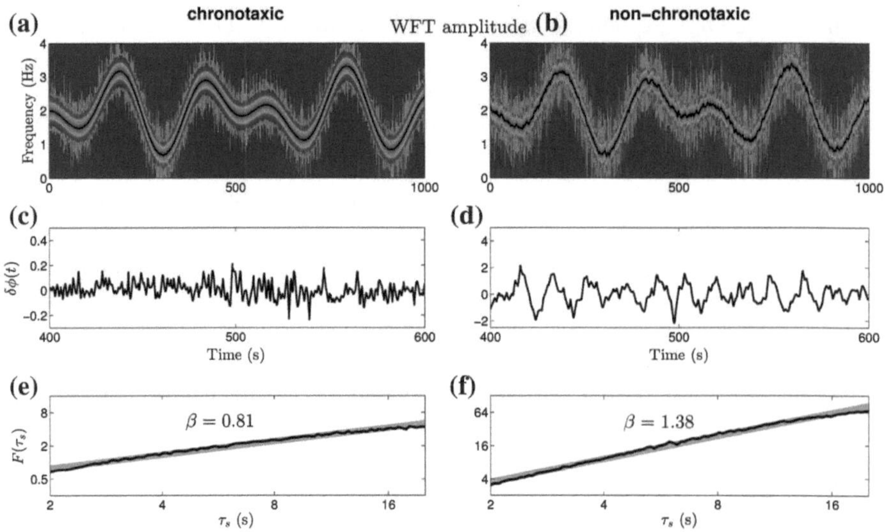

Fig. 5.15 **a**, **b** WFT amplitudes for the signals $\cos\phi_c(t)$ and $\cos\phi_{nc}(t)$, where (respectively chronotaxic and non-chronotaxic) phases $\phi_c(t)$ and $\phi_{nc}(t)$ are given by (5.11); *black lines* show the extracted *ridge curves* $\omega_p(t)$, while *thin gray lines* indicate the time-frequency supports $[\omega_-(t), \omega_+(t)]$ from which the oscillations' phases and frequencies are reconstructed by the direct method (see Sect. 3.8). **c**, **d** The phase difference $\delta\phi(t)$ (5.9), calculated with the parameters $\tau_m = \tau_d = 20$ s in (5.7), (5.9). **e**, **f** Dependence of the fluctuation strength $F(\tau_s)$ (5.10) on the scales $\tau_s \in [2, 20]$ s for $\delta\phi(t)$ shown in (**c**, **d**); in each case, the *gray background line* shows a linear fit of $\log F(\tau_s)$ with $\log \tau_s$, and β indicates its slope. Both signals are sampled at 50 Hz for 1000 s

being used is based on a particular set of assumptions, these oscillations could in principle still be chronotaxic though falling outside the model considered in [3]. Additionally, although the cardiac phase by itself does not seem to be chronotaxic, by studying the ECG signal it was shown [3] that the respiratory modulation of heart activity has pronounced chronotaxic features. It is therefore expected that, with time, chronotaxicity will be revealed in many real systems across the broad range of scientific areas.

Remark 5.3.3 It should be noted that the signal does not always reflect well the underlying physiological mechanisms that generated it. Here one of the biggest issues is whether or not the microscopic oscillations manifest themselves at the macroscopic level as a single oscillatory process [33]. Thus, for real signals one usually does not know whether the extracted mode represents a single oscillation or, instead, reflects the combined contributions of many distributed oscillators that have similar frequencies (e.g. are synchronized to some extent or driven by a common force). For example, blood flow as measured by LDF represents the combined effect of the flows of a large number of capillaries and arterioles under the probe; the more consistent the individual flow oscillations are at different locations, the more consistent and strong will be the corresponding overall oscillation in the measured flow. This issue complicates all aspects of the analysis, and especially those related to the classification of the underlying processes.

5.4 Limitations and Possible Issues

Although usually being very accurate and noise-robust, NMD does not represent an "ultimate" solution to any decomposition problem, as it has certain limitations. First of all, NMD is designed to operate within a particular set of assumptions which, although being not too restrictive, can sometimes be violated (see Sect. 5.3.1). Thus, for a reliable decomposition it should in principle be possible to represent signal as a superposition of meaningful oscillations (4.2) with amplitudes and frequencies that admit analytic approximation (2.4), i.e. vary smoothly enough.

When all underlying assumptions hold, the main restriction of the method is that it needs at least some harmonic of each mode contained in the signal to be relatively well represented in the signal's WFT or WT. Otherwise, the component extracted at the first step will not pass the surrogate test (see Sect. 4.1.5), being regarded as noise, and the procedure will stop. This might happen e.g. when the noise is exceedingly strong, so that even NMD with its exceptional noise-robustness cannot cope with it. In such circumstances one can either adjust the initial resolution parameter $f_0^{(1)}$, which might sometimes help, or continue the procedure anyway, but in this case the resultant NMs might not have any physical meaning. Furthermore, even when signal is represented by a superposition of well-defined oscillations, there exist situations when it is in principle impossible to recover reliably the modes from its TFR due to the restrictions on time and frequency resolutions of the latter (see e.g. Fig. 4.3 and related discussion); NMD will obviously be inaccurate in such a case.

Nevertheless, it appears that if NMD fails then other methods, such as (E)EMD and single-channel PCA/ICA, will not produce meaningful results as well. At the same time, NMD can perform well even when no alternative approach succeeds, as was demonstrated in this chapter (see also Sect. 1.1). Moreover, when multiple methods work, the results obtained with NMD are usually the most accurate ones.

Generally, the only noticeable disadvantage of NMD as compared to other approaches is its relatively low speed. Thus, although being $O(N \log N)$, the computational cost of the method can become considerable in some cases due to numerous recalculations of the TFRs while adapting the resolution parameters for harmonics (see Sect. 4.2.1); adjusting numerical parameters, though, can make NMD much faster at the cost of slightly decreased accuracy. Note also, that NMD addresses the problem of decomposition and filtering of a single time series, while for a large sets of signals PCA and ICA might sometimes be preferable. Finally, it should be noted that if one is interested not in all the properties of the mode, but only in its time-averaged characteristics, then simpler methods can often be used (e.g. mean amplitudes of the components can be estimated based on the areas under the corresponding peaks in the time-averaged TFR amplitude).

References

1. H.D.I. Abarbanel, R. Brown, J.J. Sidorowich, L.S. Tsimring, The analysis of observed chaotic data in physical systems. Rev. Mod. Phys. **65**(4), 1331 (1993)
2. P. Clemson, A. Stefanovska, M. Robnik, V.G. Romanovski, Time series analysis of turbulent and non-autonomous systems. In: AIP Conference Proceedings, vol. 1468 (2012), p. 69
3. P. Clemson, Y. Suprunenko, T. Stankovski, A. Stefanovska, Inverse approach to chronotaxic systems for single-variable time series. Phys. Rev. E **89**, 032904 (2014)
4. P.T. Clemson, A. Stefanovska, Discerning non-autonomous dynamics. Phys. Rep. (2014). (in press)
5. P. Comom, Independent component analysis, a new concept? Sig. Process. **36**(3), 287–314 (1994)
6. G. Duffing, Erzwungene Schwingungen bei veränderlicher Eigenfrequenz und ihre technische Bedeutung. In: Sammlung Vieweg, 41–42 (F. Vieweg & Sohn, 1918)
7. A. Duggento, T. Stankovski, P.V. McClintock, A. Stefanovska, Dynamical Bayesian inference of time-evolving interactions: from a pair of coupled oscillators to networks of oscillators. Phys. Rev. E **86**(6), 061126 (2012)
8. J.-P. Eckmann, D. Ruelle, Ergodic theory of chaos and strange attractors. Rev. Mod. Phys. **57**(3), 617 (1985)
9. F. Feihl, L. Liaudet, B. Waeber, B.I. Levy, Hypertension: a disease of the microcirculation? Hypertension **48**(6), 1012–1017 (2006)
10. M. Feldman, Analytical basics of the EMD: two harmonics decomposition. Mech. Syst. Sig. Proc. **23**(7), 2059–2071 (2009)
11. R.L. Honeycutt, Stochastic Runge-Kutta algorithms. I. White noise. Phys. Rev. A **45**(2), 600 (1992)
12. M. Hožič, A. Stefanovska, Karhunen—Loève decomposition of peripheral blood flow. Physica A **281**, 587–601 (2000)
13. N.E. Huang, Z. Shen, S.R. Long, M.C. Wu, H.H. Shih, Q. Zheng, N. Yen, C.C. Tung, H.H. Liu, The empirical mode decomposition and the Hilbert spectrum for nonlinear and non-stationary time series analysis. Proc. R. Soc. Lond. A **454**, 903–995 (1998)
14. A. Hyvärinen, E. Oja, Independent component analysis: algorithms and applications. Neural Networks **13**(4–5), 411–430 (2000)
15. D. Iatsenko, A. Bernjak, T. Stankovski, Y. Shiogai, P.J. Owen-Lynch, P.B.M. Clarkson, P.V.E. McClintock, A. Stefanovska, Evolution of cardio-respiratory interactions with age. Phil. Trans. R. Soc. Lond. A **371**(1997), 20110622 (2013)
16. D. Iatsenko, P.V.E. McClintock, A. Stefanovska, Linear and synchrosqueezed time-frequency representations revisited: overview, standards of use, reconstruction, resolution, concentration, and algorithms. Dig. Sig. Proc. (2015). doi:10.1016/j.dsp.2015.03.004. (in press)
17. N. Japundzic, M.-L. Grichois, P. Zitoun, D. Laude, J.-L. Elghozi, Spectral analysis of blood pressure and heart rate in conscious rats: effects of autonomic blockers. Auton. Nerv. Syst. **30**(2), 91–100 (1990)
18. C. Julien, The enigma of Mayer waves: facts and models. Cardiovasc. Res. **70**(1), 12–21 (2006)
19. H. Kantz, T. Schreiber, *Nonlinear Time Series Analysis* (Cambridge University Press, 2004)
20. I. Kovacic, M.J. Brennan, *The Duffing Equation: Nonlinear Oscillators and Their Behaviour* (Wiley, Chichester, 2011)

21. P. Kvandal, S.A. Landsverk, A. Bernjak, A. Stefanovska, H.D. Kvernmo, K.A. Kirkebøen, Low-frequency oscillations of the laser Doppler perfusion signal in human skin. Microvasc. Res. **72**(3), 120–127 (2006)
22. B.I. Levy, Artery changes with aging: degeneration or adaptation? Dialog. Cardiovas. Med. **6**(2), 104–111 (2001)
23. E.N. Lorenz, Deterministic nonperiodic flow. J. Atmos. Sci. **20**(2), 130–141 (1963)
24. M.B. Lotric, A. Stefanovska, D. Stajer, V. Urbancic-Rovan, Spectral components of heart rate variability determined by wavelet analysis. Physiol. Meas. **21**(4), 441 (2000)
25. M. Malik, Heart rate variability. Ann. Noninvas. Electro. **1**(2), 151–181 (1996)
26. S.C. Malpas, Neural influences on cardiovascular variability: possibilities and pitfalls. Am. J. Physiol.: Heart. Circ. Physiol. **282**, H6–H20 (2002)
27. W. Nichols, M. O'Rourke, C. Vlachopoulos, *McDonald's blood flow in arteries: theoretical, experimental and clinical principles* (CRC Press, Boca Raton, 2011)
28. G.E. Nilsson, T. Tenland, P.A. Oberg, Evaluation of a laser Doppler flowmeter for measurement of tissue blood flow. IEEE Trans. Biomed. Eng. **27**(10), 597–604 (1980)
29. C.-K. Peng, S.V. Buldyrev, S. Havlin, M. Simons, H.E. Stanley, A.L. Goldberger, Mosaic organization of DNA nucleotides. Phys. Rev. E **49**(2), 1685 (1994)
30. G. Rilling, P. Flandrin, One or two frequencies? the empirical mode decomposition answers. IEEE Trans. Sig. Proc. **56**(1), 85–95 (2008)
31. O.E. Rössler, An equation for continuous chaos. Phys. Lett. A **57**(5), 397–398 (1976)
32. J.P. Saul, R. Berger, P. Albrecht, S. Stein, M.H. Chen, R. Cohen, Transfer function analysis of the circulation: unique insights into cardiovascular regulation. Am. J. Physiol.-Heart Circulatory Physiol. **261**(4), H1231–H1245 (1991)
33. L.W. Sheppard, A.C. Hale, S. Petkoski, P.V.E. McClintock, A. Stefanovska, Characterizing an ensemble of interacting oscillators: the mean-field variability index. Phys. Rev. E **87**(1), 012905 (2013)
34. Y. Shiogai, A. Stefanovska, P.V.E. McClintock, Nonlinear dynamics of cardiovascular ageing. Phys. Rep. **488**, 51–110 (2010)
35. P. Sörös, S. Whitehead, J.D. Spence, V. Hachinski, Antihypertensive treatment can prevent stroke and cognitive decline. Nat. Revs. Neurol. **9**(3), 174–178 (2013)
36. G. Srivastava, S. Crottaz-Herbette, K.M. Lau, G.H. Glover, V. Menon, ICA-based procedures for removing ballistocardiogram artifacts from EEG data acquired in the MRI scanner. NeuroImage **24**, 50–60 (2005)
37. T. Stankovski, A. Duggento, P.V. McClintock, A. Stefanovska, Inference of time-evolving coupled dynamical systems in the presence of noise. Phys. Rev. Lett. **109**(2), 024101 (2012)
38. A. Stefanovska, M. Bračič, Physics of the human cardiovascular system. Contemp. Phys. **40**(1), 31–55 (1999)
39. A. Stefanovska, M. Bračič, H.D. Kvernmo, Wavelet analysis of oscillations in the peripheral blood circulation measured by laser Doppler technique. IEEE Trans. Bio. Med. Eng. **46**(10), 1230–1239 (1999)
40. A. Stefanovska, P.V. McClintock, J. Ræder, A. Smith, *Nonlinear Dynamics of Anaesthesia—From Theory to Clinical Practice* (Springer, 2014)
41. Y. Suprunenko, P. Clemson, A. Stefanovska, Chronotaxic systems: a new class of self-sustained non-autonomous oscillators. Phys. Rev. Lett. **111**(2), 024101 (2013)
42. Y. Suprunenko, P. Clemson, A. Stefanovska, Chronotaxic systems with separable amplitude and phase dynamics. Phys. Rev. E **89**(1), 012922 (2014)
43. S. Taddei, A. Virdis, L. Ghiadoni, I. Sudano, A. Salvetti, Antihypertensive drugs and reversing of endothelial dysfunction in hypertension. Curr. Hypertens. Rep. **2**(1), 64–70 (2000)
44. P. Verdecchia, C. Porcellati, G. Schillaci, C. Borgioni, A. Ciucci, M. Battistelli, M. Guerrieri, C. Gatteschi, I. Zampi, A. Santucci et al., Ambulatory blood pressure. An independent predictor of prognosis in essential hypertension. Hypertension **24**(6), 793–801 (1994)
45. Z. Wu, N.E. Huang, Ensemble empirical mode decomposition: a noise-assisted data analysis method. Adv. Adapt. Data Anal. **1**(1), 1–41 (2009)

Chapter 6
Conclusion

6.1 Summary

This work has introduced a new method for decomposing a given signal into a set of oscillatory modes with time-varying amplitudes and frequencies—nonlinear mode decomposition (NMD). It is based on time-frequency analysis techniques [1, 2], surrogate data tests [3–6], and the idea of identification of the time-variable harmonics [7]. Unlike many previous methods, NMD retrieves the oscillations within *any* waveform, and not only the sinusoidal one. It is also extremely noise-robust and in a sense super-adaptive, meaning that most of its settings are automatically adapted to the properties of the particular signal under investigation. Finally, in contrast to the other methods, NMD returns only the physically meaningful oscillations, stopping the decomposition when the residual is just noise (any type within a large class, see Sect. 4.1.5).

All the relevant aspects of NMD and its subprocedures (each of which represents a useful technique in its own right, see Sect. 4.1) have been thoroughly discussed. Thus, this work includes a review and, to some extent, revision of the linear time-frequency representations on which NMD is based, namely the WFT and the WT. Furthermore, the techniques for extraction of the components from these TFRs have been developed and considered in detail.

The exceptional performance of NMD has been illustrated on both simulated and real data, including some important applications. Thus, it has been shown that the method allows one to study effectively many real-life signals such as skin blood flows, electrocardiograms, electroencephalograms etc. NMD can therefore be used in clinical practice, and it has already been applied to uncover certain aspects of cardiovascular ageing and hypertension. Finally, NMD is a necessary prerequisite for many other methods, e.g. those devoted to identifying the chronotaxicity of the systems (see Sect. 5.3.2).

The area of applicability of NMD, however, is not limited to the cases considered, being much wider. For example, it would now be reasonable to reconsider all those cases where other decomposition methods, such as (E)EMD [8, 9], have been applied

D. Iatsenko, *Nonlinear Mode Decomposition*,
Springer Theses, DOI 10.1007/978-3-319-20016-3_6

(e.g. see references in [10, 11]). Furthermore, the exceptional noise-robustness of NMD and its other advantages allow one to use it even in situations where other methods fail completely. Thus, it can be applied to a majority of multicomponent signals of the kind that are ubiquitous in the life sciences, geophysics, astrophysics, econometrics etc. It is therefore expected that, with time, NMD will become a new standard in the field.

The latest MATLAB codes needed for running NMD and its individual subprocedures are freely available [12], together with detailed instructions and examples, in both text and video formats. There are a few pre-set templates controlling the speed-accuracy tradeoff, while most of the other parameters of the method are either quite universal or are selected adaptively (see Sect. 4.3), so that NMD requires minimal tuning and should be easy to apply even for inexperienced users.

6.2 Original Contributions

Basically, each Chapter of this work contains original contributions and ideas, with their concentration growing from the beginning towards the end of the work:

- In addition to a thorough review of linear TFRs and their properties, the work has also advanced their value by introducing some new procedures and by elaborating certain aspects of their use. Thus, the new contributions associated with Chap. 2 include: derivation of the general formulas for reconstructing the signal and its parameters from the corresponding WFT/WT (Sect. 7.3); introduction of the window/wavelet ϵ-support formalism (2.26), (2.27) ' time-frequency resolution measures (Sect. 2.3); the criteria (2.34) for selecting the optimal frequency bin width (Sect. 2.4.2); derivation of the rigorous expressions for the boundary errors (2.37), (2.38) and cone-of-influence (2.40), as well as for the number of padded data points (2.35), (2.36), and introduction of the predictive padding scheme (Sect. 2.4.3 and Sect. 7.4).
- This work has elaborated in detail the identification and reconstruction of the components from the signal's WFT/WT, and many powerful techniques have been developed for this task (Chap. 3).
- Nonlinear mode decomposition and its subprocedures have been methodologically developed and considered in detail (Chap. 4); some related issues, e.g. the optimal representation of harmonics (Sect. 4.2.1), have also been discussed.
- By applying NMD to real data, the work has approached successfully certain previously unsolved problems, such as the decomposition of the skin blood flow signal and the removal of cardiac artifacts from a single EEG recording. Moreover, with the help of NMD certain new aspects of cardiovascular ageing and hypertension have been revealed. Performance of the method for different systems and its possible use for system classification have also been considered.

6.3 Future Perspectives

The expected future of this research can be outlined as discussed below.

- Apart from those described in this work, there are many other possible applications of the method. Thus, it is expected that NMD will be routinely applied to the multicomponent signals that appear naturally in different areas of research, and that this work will therefore contribute to the understanding of the nature and properties of the oscillatory processes encountered in a wide range of real systems.
- Although NMD is already a very powerful tool, it might still be significantly improved by developing a scheme for the adaptive selection of the initial resolution parameter f_0 (i.e. the one that is used for extracting the first harmonic, see Sect. 4.3.1). Furthermore, the TFR-based surrogate test against noise (Sect. 4.1.5) can probably be made more powerful by selecting a better discriminating statistic, the performance of which is not influenced by the relationship between the strengths of the amplitude and frequency modulations of the component, as is the case for $D(\alpha_A, \alpha_\nu)$ (4.9). It is also possible that, instead of choosing between the direct and ridge estimates (Sect. 3.2), some universal method for reconstructing parameters of the component based on its extracted ridge curve can be devised and used.
- In the same way as NMD has been used to uncover aspects of cardiovascular ageing and hypertension, it can be widely applied in clinical studies to investigate how the properties of the blood flow (or other) signals depend on health and physiological state. For example, it will be very interesting to see how these properties change for cardiac failure and diabetic subjects.
- As discussed in Sect. 5.3.2, NMD can be combined with other signal analysis techniques, and a variety of such combinations is expected to be found.

References

1. D. Iatsenko, P.V.E. McClintock, A. Stefanovska, On the extraction of instantaneous frequencies from ridges in time-frequency representations of signals [preprint—arXiv:1310.7276], 2014
2. D. Iatsenko, P. V. E. McClintock, A. Stefanovska, Linear and synchrosqueezed time-frequency representations revisited: Overview, standards of use, reconstruction, resolution, concentration, and algorithms. Digital Signal Processing (in press), 2015, doi:10.1016/j.dsp.2015.03.004
3. R.G. Andrzejak, A. Kraskov, H. Stögbauer, F. Mormann, T. Kreuz, Bivariate surrogate techniques: necessity, strengths, and caveats. Phys. Rev. E **68**(6), 066202 (2003)
4. R.Q. Quiroga, A. Kraskov, T. Kreuz, P. Grassberger, Performance of different synchronization measures in real data: a case study on electroencephalographic signals. Phys. Rev. E **65**(4), 041903 (2002)
5. T. Schreiber, A. Schmitz, Surrogate time series. Physica D **142**(3), 346–382 (2000)
6. J. Theiler, S. Eubank, A. Longtin, B. Galdrikian, J.D Farmer, Testing for nonlinearity in time series: the method of surrogate data. Physica D **58**(1), 77–94 (1992)
7. L.W. Sheppard, A. Stefanovska, P.V.E. McClintock, Detecting the harmonics of oscillations with time-variable frequencies. Phys. Rev. E **83**, 016206 (2011)

8. N.E. Huang, Z. Shen, S.R. Long, M.C. Wu, H.H. Shih, Q. Zheng, N. Yen, C.C. Tung, H.H. Liu, The empirical mode decomposition and the Hilbert spectrum for nonlinear and non-stationary time series analysis. Proc. R. Soc. Lond. A **454**, 903–995 (1998)
9. Z. Wu, N.E. Huang, Ensemble empirical mode decomposition: a noise-assisted data analysis method. Adv. Adapt. Data Anal. **1**(1), 1–41 (2009)
10. N.E. Huang, S.S. Shen, *Hilbert-Huang Transform and its Applications* (World Scientific, Singapore 2005)
11. N.E. Huang Z. Wu, A review on Hilbert-Huang transform: method and its applications to geophysical studies. Rev. Geophys. **46**(2), RG2006 (2008)
12. The NMD Toolbox for MatLab, which contains all the needed codes, is freely available at http://www.physics.lancs.ac.uk/research/nbmphysics/diats/nmd/ (see also http://www.physics.lancs.ac.uk/research/nbmphysics/diats/tfr/)

Chapter 7
Appendix: Useful Information and Derivations

7.1 Errors of the Analytic Estimates

For a single AM/FM component (2.1) the error of the analytic estimates $A^a(t)$, $\phi^a(t)$ (2.4) of its amplitude and phase $A(t)$, $\phi(t)$ can be quantified by $\varepsilon^a(t) \equiv A(t)e^{i\phi(t)} - A^a(t)e^{i\phi^a(t)}$. Substituting $s(t) = (1/2)[A(t)e^{i\phi(t)} + (A(t)e^{i\phi(t)})^*]$ into (2.3), and taking into account that for real signals $\widehat{s^*}(\xi) = [\hat{s}(-\xi)]^*$, one obtains

$$s^a(t) \equiv A^a(t)e^{i\phi^a(t)} = [A(t)e^{i\phi(t)}]^+ + \left([A(t)e^{i\phi(t)}]^-\right)^*, \tag{7.1}$$

Then, subtracting from $A(t)e^{i\phi(t)} = \langle A(t)e^{i\phi(t)}\rangle + [A(t)e^{i\phi(t)}]^+ + [A(t)e^{i\phi(t)}]^-$ the expression for $A^a(t)e^{i\phi^a(t)}$ (7.1), the error of the analytic approximation (2.4) can be brought to a form

$$\varepsilon^a(t) \equiv A(t)e^{i\phi(t)} - A^a(t)e^{i\phi^a(t)} = \langle Ae^{i\phi(t)}\rangle + 2i\,\mathrm{Im}\left[\left[A(t)e^{i\phi(t)}\right]^-\right]. \tag{7.2}$$

It is usually dominated by the second (purely imaginary) term, while $\langle Ae^{i\phi(t)}\rangle$ is often negligible or is exactly zero (see below). Note, that by differentiating (7.2) one can also find the inaccuracies of the analytic estimates for the higher amplitude/phase time-derivatives, e.g. that for the analytic frequency $\partial_t\phi(t) - \partial_t\phi^a(t) = O(\partial_t\varepsilon^a(t), \varepsilon^a(t))$.

To better understand the properties of $\varepsilon^a(t)$ (7.2), consider an AM/FM component with a simple sinusoidal law of amplitude and frequency modulation:

$$\begin{aligned} s(t) &= (1 + r_a\cos[\nu_a t + \varphi_a])\cos[\nu t + \varphi + r_b\sin(\nu_b t + \varphi_b)] \\ &= \frac{1}{2}\left[\left(1 + \frac{1}{2}r_a e^{i(\nu_a t+\varphi_a)} + \frac{1}{2}r_a e^{-i(\nu_a t+\varphi_a)}\right)\sum_{n=-\infty}^{\infty} J_n(r_b)e^{i[(\nu+n\nu_b)t+(\varphi+n\varphi_b)]}\right] + c.c., \end{aligned} \tag{7.3}$$

where the expansion $e^{ia\sin\phi} = \sum_{n=-\infty}^{\infty} J_n(a)e^{in\phi}$ was used, $J_n(x) = (-1)^n J_{-n}(x)$ denote Bessel functions of the first kind, and $c.c.$ stands for the complex conjugate of the preceding expression. Note that, because the definition of phase implies $\phi'(t) \geq 0$, one has a restriction $r_b\nu_b \leq \nu$ in (7.4), whereas according to the amplitude definition $A(t) \geq 0$ one has $r_a \leq 1$.

D. Iatsenko, *Nonlinear Mode Decomposition*,
Springer Theses, DOI 10.1007/978-3-319-20016-3_7

Substituting (7.3) into (7.2), one obtains

$$\begin{aligned}
\varepsilon^a(t) &= \frac{1}{2}\langle\widetilde{\varepsilon}(t)\rangle + i\mathrm{Im}[\widetilde{\varepsilon}(t)], \\
\widetilde{\varepsilon}(t) &\equiv \sum_{n:\nu+n\nu_b\le 0} J_n(r_b)e^{i(\varphi+n\varphi_b)}e^{i(\nu+n\nu_b)t} \\
&+ \frac{r_a}{2}\sum_{n:\nu+n\nu_b+\nu_a\le 0} J_n(r_b)e^{i(\varphi+n\varphi_b+\varphi_a)}e^{i(\nu+n\nu_b+\nu_a)t} \\
&+ \frac{r_a}{2}\sum_{n:\nu+n\nu_b-\nu_a\le 0} J_n(r_b)e^{i(\varphi+n\varphi_b-\varphi_a)}e^{i(\nu+n\nu_b-\nu_a)t}.
\end{aligned} \tag{7.4}$$

From (7.4) it can be seen, that the average $\langle A(t)e^{i\phi(t)}\rangle$ is not zero only when $\nu + n\nu_b = 0$ or $\nu + n\nu_b \pm \nu_a = 0$ for some n, i.e. when the frequency content of $A(t)$ intersects with the frequency content of $e^{i\phi(t)}$. Obviously, this is rather a special case, which requires particular relationships between ν, ν_a, ν_b.

As can be seen from (7.4), the error $\varepsilon^a(t)$ is proportional to r_a, r_b, but the coefficients of this proportionality are determined by the values of n for which $\nu+n\nu_b \le 0$ and/or $\nu+n\nu_b\pm\nu_a \le 0$. Hence, the behavior of $\varepsilon^a(t)$ can be partitioned into different regimes, separated by discontinuous step increases in error. When only amplitude, but not frequency, modulation is present ($r_b = 0 \Rightarrow J_{n\neq 0}(r_b) = 0$, so that one can set $n = 0$ in all terms of (7.4)), it follows that the analytic approximation is exact ($\varepsilon^a(t) = 0$) when $\nu_a < \nu$; otherwise the error is linearly proportional to r_a. When there is no amplitude modulation ($r_a = 0$), but frequency modulation exists, the $\varepsilon^a(t)$ depends on r_b and ν_b in a complicated fashion, as the relationships between $J_n(r_b)$ in (7.4) change with r_b. However, when $r_b \in [0, 1]$, one has $|J_{n_1}(r_b)| \gg |J_{n_2}(r_b)|$ if $|n_1| < |n_2|$. Therefore, in this case the error is largely determined by the smallest n for which $\nu - n\nu_b \le 0$, being negligible for large n but often considerable for $n \le 2$; it is also proportional to r_b, but the dependence is not simple.

The latter case is illustrated in Fig. 7.1a, where it can be seen that the relative error of the analytic approximation increases in steps when ν_b passes the levels ν/n. Because for $r_b \le 1$ one has $J_{n>3}(r_b) < 0.0025$, it becomes non-negligible only for $\nu_b \ge \nu/3$. In the case when both amplitude and frequency modulations are present the situation becomes more complicated, on account of an additional contribution from the mixing terms $\nu + n\nu_b \pm \nu_a \le 0$ in (7.4). Thus, there appear additional step increases in error when ν_b crosses not only the levels ν/n, but also $(\nu \pm \nu_a)/n$, as shown in Fig. 7.1b–d. Nevertheless, for $r_b \in [0, 1]$, not-small r_a and $\nu_a < \nu$, the error is largely determined by the smallest n for which $\nu - \nu_a - n\nu_b \le 0$, as is clearly seen in Fig. 7.1d.

Amplitude/frequency modulations more complex than (7.3) can always be expanded in Fourier series (see 2.6), so that the expression for $\varepsilon^a(t)$ (7.4) in this case will include additional terms

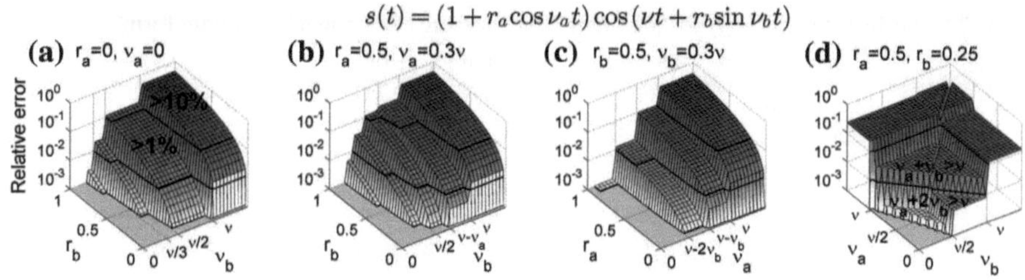

Fig. 7.1 Relative error of the analytic approximation $\langle|\varepsilon^a(t)|^2\rangle/\langle A^2(t)\rangle$, where $\varepsilon^a(t)$ is given by (7.2), in dependence on signal parameters (assuming infinite time-length and sampling frequency): **a** its dependence on the parameters of frequency modulation r_b, ν_b for the FM component $s(t) = \cos(\nu t + r_b \sin \nu_b t)$; **b** its dependence on the parameters of frequency modulation r_b, ν_b for the AM/FM component $s(t) = (1 + 0.5\cos(0.3\nu t))\cos(\nu t + r_b \sin \nu_b t)$, with parameters of the amplitude modulation being fixed; **c** its dependence on the parameters of amplitude modulation r_a, ν_a for the AM/FM component $s(t) = (1 + r_a \cos(\nu_a t))\cos(\nu t + 0.5 \sin 0.3\nu t)$, with parameters of the frequency modulation being fixed; **d** its dependence on the frequencies of amplitude and frequency modulations ν_a, ν_b for the AM/FM component $s(t) = (1 + 0.5\cos(\nu_a t))\cos(\nu t + 0.25 \sin \nu_b t)$, with other parameters remaining fixed. *Solid black lines* indicate levels of 0.001, 0.01 and 0.1; for simplicity, the behavior of the relative error below 0.001 is not shown

$$\sim r_a^{(i)} \sum_{\{n^{(j)}\}:\nu \pm \nu_a^{(i)} + \sum_j n^{(j)}\nu_b^{(j)} \le 0} e^{i(\nu \pm \nu_a^{(i)} + \sum_j n^{(j)}\nu_b^{(j)})t} \prod_j J_{n^{(j)}}(r_b^{(j)})$$

for each Fourier term i in the amplitude modulation (including the DC value $\nu_a^{(0)} = 0$) and each Fourier term j in the frequency modulation; the behavior of the error will therefore be partitioned into additional possible regimes.

7.2 Window and Wavelet Functions and Their Properties

This Appendix discusses the properties of the most common window and wavelet functions. All of them are implemented in the codes [1] under the specific names, which are provided in the tables below, though one can use any window/wavelet function by specifying either its frequency domain or time domain form (or both, if known).

7.2.1 *Window Functions*

Table 7.1 lists the commonest window forms and related quantities. The resolution characteristics (2.32) for each window are presented in Fig. 7.2a–c. For completeness, the "classic" resolution characteristics (2.24) are also shown in Fig. 7.2d–f, although

Table 7.1 Different window types and their characteristics (if known in analytic form)

Name	Name in codes	Description and characteristics
Gaussian	"Gaussian" (default)	$g(t) = \frac{1}{\sqrt{2\pi} f_0} e^{-(t/f_0)^2/2},\ t \in (-\infty, \infty)$, $\hat{g}(\xi) = e^{-(f_0\xi)^2/2},\ \xi \in (-\infty, \infty)$, $R_g(\omega) = \frac{1}{2}\left[\mathrm{erf}\left(f_0^{-1}\omega/\sqrt{2}\right) + 1\right],\ \xi_{1,2}(\epsilon) = \mp f_0^{-1} n_G(\epsilon)$, $P_g(\tau) = \frac{1}{2}\left[\mathrm{erf}\left(f_0\tau/\sqrt{2}\right) + 1\right],\ \tau_{1,2}(\epsilon) = \mp f_0 n_G(\epsilon)$, $C_g = \sqrt{\frac{\pi}{2}} f_0^{-1},\ \overline{\omega}_g = 0$.
Hann	"Hann"	$q = 4.4 f_0$, $g(t) = (1 + \cos(2\pi t/q))/2 = \sin^2(\pi t/q + \pi/2),\ t \in (-q/2, q/2)$, $\hat{g}(\xi) = \frac{-(4\pi^2/q^2)\sin(\xi q/2)}{\xi(\xi^2 - 4\pi^2/q^2)},\ \xi \in (-\infty, \infty)$, $P_g(\tau) = 2\tau/q - (1/2\pi)\sin(2\pi\tau/q)$, $C_g = \pi,\ \overline{\omega}_g = 0$.
Blackman	"Blackman"	$q = 5.6 f_0,\ \alpha = 0.16$, $g(t) = (1 + \cos(2\pi t/q))/2 - \alpha(1 + \cos(4\pi t/q))/2,\ t \in (-q/2, q/2)$, $\hat{g}(\xi) = \frac{-(4\pi^2/q^2)\sin(\xi q/2)}{\xi}\left[\frac{1}{\xi^2 - 4\pi^2/q^2} - \frac{4\alpha}{\xi^2 - 16\pi^2/q^2}\right],\ \xi \in (-\infty, \infty)$, $P_g(\tau) = 2\tau/q - [(1/2\pi)\sin(2\pi\tau/q) - (\alpha/4\pi)\sin(4\pi\tau/q)]/(1-\alpha)$, $C_g = \pi(1-\alpha),\ \overline{\omega}_g = 0$.
Exponential	"Exp"	$q = 6.5 f_0$, $g(t) = e^{-\lvert t\rvert/q},\ t \in (-\infty, \infty)$, $\hat{g}(\xi) = 2q^{-1}[\xi^2 + q^{-2}]^{-1},\ \xi \in (-\infty, \infty)$, $R_g(\omega) = 1/2 + \pi^{-1}\arctan(q\omega),\ \xi_{1,2}(\epsilon) = \mp q^{-1}\mathrm{ctg}(\pi\epsilon/2)$, $P_g(\tau) = 1/2 + (\mathrm{sign}(\tau)/2)[1 - e^{-q^{-1}\lvert\tau\rvert}],\ \tau_{1,2}(\epsilon) = \pm q\log\epsilon$, $C_g = \pi,\ \overline{\omega}_g = 0$.
Rectangular	"Rect"	$q = 10 f_0$, $g(t) = 1,\ t \in [-q/2, q/2]$, $\hat{g}(\xi) = 2\frac{\sin(q\xi/2)}{\xi},\ \xi \in (-\infty, \infty)$, $P_g(\tau) = \tau/q + 0.5,\ \tau_{1,2}(\epsilon) = \pm q(1-\epsilon)/2$, $C_g = \pi,\ \overline{\omega}_g = 0$.
Kaiser	"Kaiser-a" (e.g. "Kaiser-2.5")	$q = 3\sqrt{1 + \lvert a - 1/a\rvert} f_0$, $g(t) = I_0(\pi a\sqrt{1 - (2t/q)^2})/I_0(\pi a),\ t \in (-q/2, q/2)$, $C_g = \pi,\ \overline{\omega}_g = 0$.

The names under which these windows are implemented in the codes [1] are given in the second column, but one can specify any window function there. The resolution parameter f_0 for each window is adjusted in such a way that, for the same f_0, all of them have similar frequency resolutions (as defined in (2.32) with $\epsilon_r = 0.05$), see Fig. 7.2a below. Note, however, that it is hard to ensure this for all a in the Kaiser window, so some deviations are possible in that case. See List of Abbreviations and Symbols for an explanation of the mathematical notation

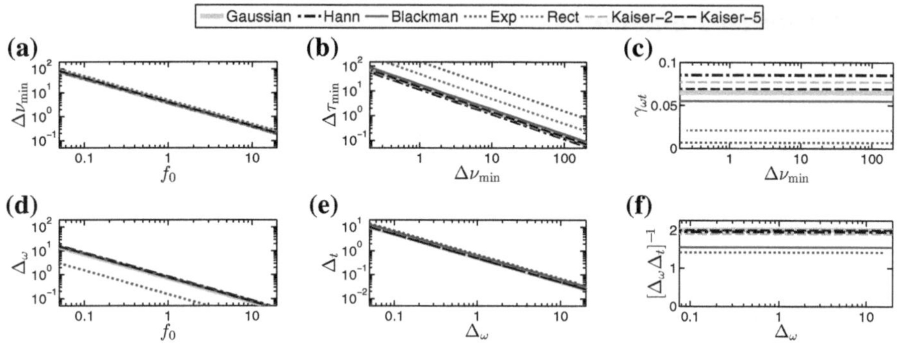

Fig. 7.2 **a–c** Resolution characteristics of the windows listed in Table 7.1, according to (2.32) with $\epsilon_r = 0.05$. The minimum resolvable frequency $\Delta\nu_{\min}$ in dependence on the window resolution parameter f_0 is shown in (**a**), while the dependences of the minimum resolvable time lag $\Delta\tau_{\min}$ and the joint time-frequency resolution $\gamma_{\omega t}$ on $\Delta\nu_{\min}$ are shown in (**b**) and (**c**), respectively. **d–f** The same as (**a–c**), but for the "classic" resolution characteristics (2.24), though they are not fully appropriate (see Sect. 2.3.2); there are no lines for the rectangular window here, as it is characterized by $\Delta_\omega = \infty$

they do not have a straightforward relationship to the resolution of two components in the WFT, as discussed in Sect. 2.3.2.

Defined in any way, the time and frequency resolutions of all windows have simple linear proportionality to f_0^{-1} and f_0, respectively (see Fig. 7.2); this is an obvious result given that in all cases $g[f_0](t) = g[1](t/f_0)$. The joint time-frequency resolutions, which remain fixed, provide more useful information. Thus, it is clear that, for all conventions, exponential and rectangular windows have much worse resolution properties compared to other windows considered. Surprisingly, in terms of time-frequency resolution the Kaiser and Hann windows outperform slightly the Gaussian window (Fig. 7.2c), which is widely believed to have the best resolution properties; these two windows are also very close to Gaussian even in terms of the (not fully appropriate) "classic" $\gamma_{\omega t}$ (Fig. 7.2f), that is maximized for the latter [2, 10, 13].

However, the Kaiser and Hann windows, apart from worse analytical tractability, have one very significant drawback as compared to the Gaussian window: they are not unimodal in the frequency domain, which makes the time-frequency support (see Sect. 3.2.2) ill-defined and the reconstruction of components from the WFT more problematic. Because the corresponding difference in the time-frequency resolution is not huge (Fig. 7.2c), the Gaussian window therefore remains a preferred choice and is implemented as default in the codes.

7.2.2 Wavelet Functions

Table 7.2 lists the commonest wavelet forms and related quantities. Note that many wavelets are included within the generalized Morse family, which was introduced in [14] and studied in detail in [11, 12]. Thus, this family includes Cauchy ($a = 1$), the derivative of Gaussian ($a = 2$), Airy ($a = 3$) and other wavelets [12]. Different resolution characteristics for each wavelet are presented in Fig. 7.3a–c. For completeness, the "classic" resolution characteristics (2.25) are also shown in Fig. 7.3d–f but, as discussed in Sect. 2.3.2, they are actually irrelevant and therefore will not be considered in what follows.

From Fig. 7.3a, b one can see that the widely used Morlet wavelet has the upper (lower) limit on its time (frequency) resolution. Thus, for this wavelet type, $\Delta\nu_{\min}(2\pi)$ saturates at some level when f_0 decreases, and the same happens with $\Delta\tau_{\min}(2\pi)$. As a result, lowering f_0 below ≈ 0.05 does not effectively change anything: neither the time, nor the frequency, nor the joint time-frequency resolution. This happens because the Morlet wavelet for $f_0 \ll 1$ becomes $\hat{\psi}(\xi) \approx 2\pi f_0 \xi e^{-\xi^2/2}$ with $\omega_\psi \approx 1$, so that f_0 determines only the (unimportant) constant multiplier.

Table 7.2 Different wavelet types and their characteristics (if known in analytic form)

Name	Name in codes	Description and characteristics
Lognormal	"Lognorm" (default)	$\hat{\psi}(\xi) = e^{-(2\pi f_0 \log\xi)^2/2},\ \xi \in (0, \infty)$,
		$R_\psi(\omega) = \frac{1}{2}\left[\operatorname{erf}\left((2\pi f_0)^{-1} \log\omega/\sqrt{2}\right) + 1\right],\ \xi_{1,2}(\epsilon) = \exp\left[\mp\frac{n_G(\epsilon)}{2\pi f_0}\right]$,
		$\omega_\psi = 1, C_\psi = \sqrt{\frac{\pi}{2}} f_0^{-1}/2\pi,\ D_\psi = C_\psi e^{\frac{1}{2}(4\pi^2 f_0^2)^{-1}}$.
Morlet	"Morlet"	$\hat{\psi}(\xi) = e^{-(\xi-2\pi f_0)^2}(1 - e^{-2\pi f_0 \xi}),\ \xi \in (0, \infty)$,
		$\psi(t) = \frac{1}{\sqrt{2\pi}} e^{-t^2/2} e^{i2\pi f_0 t} + O(e^{-(2\pi f_0)^2/2}),\ t \in (-\infty, \infty)$,
		$\omega_\psi = 2\pi f_0 + O(e^{-(2\pi f_0)^2/2}),\ D_\psi = \infty$.
Bump	"Bump"	$\Delta = 0.4 f_0^{-1} \leq 1$ (so that $f_0 \geq 0.4$),
		$\hat{\psi}(\xi) = \exp\left(1 - \frac{1}{1-\Delta^{-2}(1-\xi)^2}\right),\ \xi \in (1-\Delta, 1+\Delta)$,
		$\omega_\psi = 1,\ D_\psi < \infty$.
Generalized Morse family	"Morse-a" (e.g. "Morse-2.5")	$q = 30 f_0/a$,
		$\hat{\psi}(\xi) = B\xi^q e^{-\xi^a} = e^{-\xi^a + q\log\xi + \log B},\ \xi \in (0, \infty),\ B \equiv (ea/q)^{q/a}$,
		$D_\psi = \frac{\omega_\psi B}{2a}\Gamma((q-1)/a)$ ($= \infty$ for $q \leq 1$),
		$\omega_\psi = (q/a)^{1/a},\ C_\psi = \frac{B}{2a}\Gamma(q/a)$.

The names under which these wavelets are implemented in the codes [1] are given in the second column, but one can specify any wavelet function there. The resolution parameter f_0 for each wavelet is adjusted in such a way that at $f_0 = 1$ all of them have similar frequency resolutions (as defined in (2.33) with $\epsilon_r = 0.05$), see Fig. 7.3a below. See List of Abbreviations and Symbols for an explanation of the mathematical notation

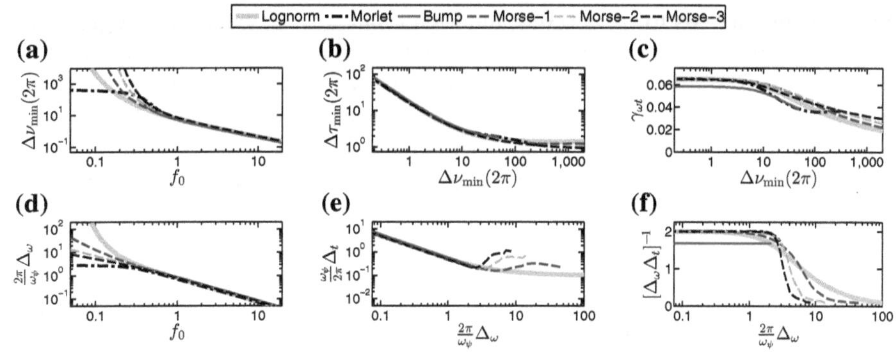

Fig. 7.3 **a–c** Resolution characteristics of the wavelets listed in Table 7.2, according to (2.33) with $\epsilon_r = 0.05$. The minimal resolvable frequency $\Delta\nu_{\min}(\omega)$ at $\omega = 2\pi$ in dependence on the wavelet resolution parameter f_0 is shown in (**a**), while the dependences of the minimum resolvable time lag $\Delta\tau_{\min}(2\pi)$ and the joint time-frequency resolution $\gamma_{\omega t}$ on $\Delta\nu_{\min}(2\pi)$ are shown in (**b**) and (**c**), respectively. **d–f** The same as (**a–c**), but for the "classic" resolution characteristics (2.25). The latter, however, are completely inappropriate, see Sect. 2.3.2

However, the impossibility of achieving high time resolution, as well as the worsening of joint time-frequency resolution for low f_0, appears to be a general property of wavelets. It is probably related to the fact that the WT effectively considers the wavelet on a linear scale in time (although rescaled at each frequency), and on a logarithmic scale in frequency (totally independent of time), as can be seen e.g. from (2.20) and (2.21). Thus, the properties of $\hat{\psi}(\xi)$ on both linear and logarithmic scales are important: the former determines the time resolution (since $\psi(t)$ is the inverse FT of the $\hat{\psi}(\xi)$, considered on a linear scale), while the latter is responsible for the frequency resolution (see Sect. 2.3).

Consider a few examples. Assuming for simplicity that the wavelet peak frequency ω_ψ is fixed, it is clear that to decrease the effective spread of the wavelet in time (and therefore increase its time resolution), one needs to increase its spread in frequency. If $\hat{\psi}(\xi)$ has a finite support in terms of $\log\xi$, then one can increase its spread around fixed ω_ψ only up to some limiting point determined by preserving the admissibility $\hat{\psi}(0) = 0$; this is the case of the bump wavelet. On the other hand, if $\hat{\psi}(\xi)$ has an infinite support in terms of $\log\xi$, then increasing its spread will typically lead to increase of the asymmetry of $\hat{\psi}(\xi)$ on a linear scale and its more rapid drop to zero as $\xi \to 0$. Both of these have negative effects on the decay of $\psi(t)e^{-i\omega_\psi t}$, thus counteracting the desired increase in time resolution; this is the case of e.g. the lognormal wavelet. Hence, whether it is possible to achieve high time resolution with the WT, and whether there exists a wavelet form for which the joint time-frequency resolution is relatively unaffected by changing the tradeoff between its time and frequency resolutions, are open questions.

From Fig. 7.3 it can be seen that the Morse wavelets have slightly better resolution properties than the other forms (Fig. 7.3c). However, with increasing ϵ_r in (2.33)

(Fig. 7.3 corresponds to $\epsilon_r = 0.05$), the lognormal wavelet becomes progressively better in terms of the time-frequency resolution $\gamma_{\omega t}$ (not shown), outperforming Morse wavelets at $\epsilon_r \gtrsim 0.1$. It also possesses the many other advantages discussed in detail in Sect. 2.2.2, thus being a preferred choice among those listed in Table 7.2 (and so the lognormal wavelet is implemented as default in the codes). Nevertheless, it remains possible that some better wavelet form could be constructed that would greatly outperform those considered.

7.3 Derivation of the Direct Reconstruction Formulas

Using the frequency domain forms of the WFT (2.8) and WT (2.13) one can show that

$$\begin{aligned}
\int \omega^n G_s(\omega,t)d\omega &= \int \omega^n d\omega \frac{1}{2\pi}\int_0^\infty \hat{s}(\xi)\hat{g}(\omega-\xi)e^{i\xi t}d\xi \\
&= /\omega = \widetilde{\omega}+\xi,\ (\widetilde{\omega}+\xi)^n = \sum_{k=0}^{n} C_n^k \widetilde{\omega}^{n-k}\xi^k/ \\
&= \sum_{k=0}^{n} C_k^n \Big(\int \widetilde{\omega}^{n-k}\hat{g}(\widetilde{\omega})d\widetilde{\omega}\Big)\Big(\frac{1}{2\pi}\int_0^\infty \xi^k \hat{s}(\xi)e^{i\xi t}d\xi\Big) \\
&= \sum_{k=0}^{n} C_k^n \Big(\frac{1}{2}\int \widetilde{\omega}^{n-k}\hat{g}(\widetilde{\omega})d\widetilde{\omega}\Big)(-i)^k\partial_t^k s^a(t)
\end{aligned} \tag{7.5}$$

$$\begin{aligned}
\int_0^\infty \omega W_s(\omega,t)\frac{d\omega}{\omega} &= \int_0^\infty \omega^{n-1}d\omega \frac{1}{2\pi}\int_0^\infty \hat{s}(\xi)\hat{\psi}^*(\omega_\psi\xi/\omega)d\xi \\
&= /\omega = \omega_\psi\xi/\widetilde{\omega}, \text{ after which the integrals decouple}/ \\
&= \Big(\omega_\psi^n \int_0^\infty \hat{\psi}^*(\widetilde{\omega})\frac{d\widetilde{\omega}}{\widetilde{\omega}^{n+1}}\Big)\Big(\frac{1}{2\pi}\int_0^\infty \xi^n \hat{s}(\xi)e^{i\xi t}d\xi\Big) \\
&= \Big(\frac{\omega_\psi^n}{2}\int_0^\infty \hat{\psi}^*(\widetilde{\omega})\frac{d\widetilde{\omega}}{\widetilde{\omega}^{n+1}}\Big)(-i)^n\partial_t^n s^a(t),
\end{aligned} \tag{7.6}$$

where it was taken into account that $\int_0^\infty \xi^k \hat{s}(\xi)e^{i\xi t}d\xi = (-i\partial_t)^k \int_0^\infty \hat{s}(\xi)e^{i\xi t}d\xi = (-i)^k\partial_t^k s^a(t)/2$, and $C_k^n \equiv \frac{n!}{k!(n-k)!}$ are the binomial coefficients.

In a similar manner, using the time domain forms of the WFT (2.8) and WT (2.13) it can be shown that

$$\begin{aligned}
\int t^n G_s(\omega,t)e^{-i\omega t}dt &= \int t^n dt \int s^+(\tau)g(\tau-t)e^{-i\omega\tau}d\tau \\
&= /t = \tau - \tilde{t},\ (\tau-\tilde{t})^n = \sum_{k=0}^{n} C_n^k (-\tilde{t})^{n-k}\tau^k / \\
&= \sum_{k=0}^{n} C_k^n \Big(\int (-\tilde{t})^{n-k} g(\tilde{t})d\tilde{t}\Big)\Big(\int \tau^k s^+(\tau)e^{-i\omega\tau}d\tau\Big) \\
&= \begin{cases} \sum_{k=0}^{n} (-1)^n C_k^n \Big(\int \tilde{t}^{n-k} g(\tilde{t})d\tilde{t}\Big)(-i)^k \partial_\omega^k \hat{s}(\omega) \ \text{if}\ \omega > 0 \\ 0 \ \text{if}\ \omega < 0 \end{cases}
\end{aligned} \tag{7.7}$$

$$\begin{aligned}
\int t^n W_s(\omega,t)e^{-i\omega t}dt &= \int t^n dt \int s^+(\tau)\psi^*(\omega(\tau-t)/\omega_\psi)\frac{\omega d\tau}{\omega_\psi} \\
&= /t = \tau - \omega_\psi\tilde{t}/\omega,\ (\tau-\omega_\psi\tilde{t}/\omega)^n = \sum_{k=0}^{n} C_n^k (-\omega_\psi\tilde{t}/\omega)^{n-k}\tau^k / \\
&= \sum_{k=0}^{n} C_k^n \Big(\frac{\omega_\psi^{n-k}}{\omega^{n-k}} \int (-\tilde{t})^{n-k} \psi^*(\tilde{t})e^{i\omega_\psi\tilde{t}}d\tilde{t}\Big)\Big(\int \tau^k s^+(\tau)e^{-i\omega\tau}d\tau\Big) \\
&= \begin{cases} \sum_{k=0}^{n} (-1)^n C_k^n \Big(\int \tilde{t}^{n-k} \psi^*(\tilde{t})e^{i\omega_\psi\tilde{t}}d\tilde{t}\Big)(-i)^k \partial_\omega^k \hat{s}(\omega) \ \text{if}\ \omega > 0 \\ 0 \ \text{if}\ \omega < 0 \end{cases}
\end{aligned} \tag{7.8}$$

where it was taken into account that $\int \tau^k s^+(\tau)e^{-i\omega\tau}d\tau = (-i\partial_\omega)^k \int s^+(\tau)e^{-i\omega\tau} d\tau = (-i)^k \partial_\omega^k \hat{s}(\omega)$ if $\omega > 0$ and $= 0$ otherwise (as only the positive frequency part of the signal $s^+(t)$ is used).

The expressions (7.5–7.8) provide a way to reconstruct any order derivatives of the signal's representation in time and frequency from its WFT/WT. Thus, the basic reconstruction formulas—(2.11) for the WFT and (2.17) for the WT—follow directly from (7.5), (7.7) and (7.6), (7.8) with $n = 0$, respectively.

The formulas for the direct frequency estimation (3.8) follow from (7.5) and (7.6) with $n = 1$, that can be rewritten as

$$\int \omega G_s(t,\omega)d\omega = C_g\overline{\omega}_g s^a(t) + C_g(-i\partial_t s^a(t)),\ \overline{\omega}_g \equiv \frac{1}{2C_g}\int \omega\hat{g}(\omega)d\omega, \tag{7.9}$$

$$\int \omega W_s(t,\omega)\frac{d\omega}{\omega} = D_\psi(-i\partial_t s^a(t)),\ D_\psi \equiv \frac{\omega_\psi}{2}\int_0^\infty \hat{\psi}^*(\omega)\frac{d\omega}{\omega^2}, \tag{7.10}$$

where $C_{g,\psi}$ are as defined in (2.11), (2.17). For a single AM/FM component (2.1) one has $s^a(t) = A^a(t)e^{i\phi^a(t)}$ and $\partial_t s^a(t) = s^a(t)\big[\frac{\partial_t A^a(t)}{A^a(t)} + i\partial_t\phi(t)\big]$, so that $\frac{\partial_t A^a(t)}{A^a(t)} = \mathrm{Re}\frac{\partial_t s^a(t)}{s^a(t)}$, $\partial_t\phi^a(t) = \mathrm{Im}\frac{\partial_t s^a(t)}{s^a(t)}$. Therefore, dividing both sides of (7.9) by $\int G_s(\omega,t)d\omega = C_g s^a(t)$ (2.11) and separating real and imaginary parts of the result, one obtains

$$\begin{aligned}\nu^a(t) \equiv \partial_t\phi^a(t) =& \mathrm{Re}\left[\frac{\int \omega G_s(\omega,t)d\omega}{\int G_s(\omega,t)d\omega} - \overline{\omega}_g\right],\\ \frac{\partial_t A^a(t)}{A^a(t)} =& -\mathrm{Im}\left[\frac{\int \omega G_s(\omega,t)d\omega}{\int G_s(\omega,t)d\omega} - \overline{\omega}_g\right].\end{aligned} \tag{7.11}$$

If the analytic amplitude and phase are the same as the actual ones (2.4), as is assumed in this work, the first of (7.11) coincides with the WFT-based direct frequency estimate (3.8), while the second one provides a way to estimate the time-derivative of the amplitude. The WT-based counterpart of (7.11) is derived in the same way, i.e. dividing both sides of (7.10) by $\int_0^\infty W_s(\omega,t)\frac{d\omega}{\omega} = C_\psi s^a(t)$ (2.17) and separating the real and imaginary parts of the result, which gives

$$\begin{aligned}\nu^a(t) \equiv \partial_t\phi^a(t) =& \mathrm{Re}\left[\frac{D_\psi^{-1}\int \omega W_s(\omega,t)\frac{d\omega}{\omega}}{C_\psi^{-1}\int W_s(\omega,t)\frac{d\omega}{\omega}}\right],\\ \frac{\partial_t A^a(t)}{A^a(t)} =& -\mathrm{Im}\left[\frac{D_\psi^{-1}\int \omega W_s(\omega,t)\frac{d\omega}{\omega}}{C_\psi^{-1}\int W_s(\omega,t)\frac{d\omega}{\omega}}\right].\end{aligned} \tag{7.12}$$

Obviously, using (7.5) and (7.6) with an appropriate n, one can derive the reconstruction formulas for higher-order derivatives $\partial_t^n A(t)$ and $\partial_t^n\phi(t)$ as well. The maximum order of n that one can go to is determined by the condition $\int \omega^n \hat{g}(\omega)d\omega < \infty$ for the WFT, and by $\int_0^\infty \hat{\psi}^*(\omega)\frac{d\omega}{\omega^{n+1}} < \infty$ for the WT. For the WFT with a Gaussian window (2.12) and for the WT with a lognormal wavelet (2.19) this condition is satisfied for all $n \geq 0$, while for the WT with a Morlet wavelet (2.18) one has $n < 1$ (so that even the instantaneous frequency cannot be estimated by (3.8), and one needs to use hybrid reconstruction (3.10) instead).

7.4 Forecasting Model for Predictive Padding

The predictive padding strategy, introduced in Sect. 2.4.3, aims to eliminate boundary effects in the signal's TFR by complementing the signal with its inferred/forecast past/future values. This appendix discusses the scheme used to predict the signal's behavior beyond its time limits.

Given a signal $s(t_n = (n-1)\Delta t)$, $n = 1,\dots,N$, consider first its extension for $t > T$, with $T = (N-1)\Delta t$ denoting its time duration. In the context of time-frequency analysis, it seems most appropriate to forecast the signal based on its spectral content. The simplest way of doing this is to represent signal as a sum of tones and continue this behavior to $t > T$. Thus, the signal is modelled as

$$s(t_n) = x(t_n) + \sigma\eta_W(t_n) = c_0 + \sum_{m=1}^{M} c_m \cos(\omega_m t_n + \varphi_m) + \sigma\eta_W(t_n)$$

$$= a_0 + \sum_{m=1}^{M} [a_m \cos\omega_m t_n + b_m \sin\omega_m t_n] + \sigma\eta_W(t_n), \tag{7.13}$$

where M denotes the chosen model order, $\eta_W(t_n)$ is Gaussian white noise of unit variance, and $x(t_n)$ stands for the noise-free signal, given by the sum of sinusoids. Having found appropriate c_m, φ_m, ω_m, the signal is then padded for $t > T$ with values of $x(t)$.

However, it is not easy to find the parameters in (7.13). It might seem at first glance that an approximation to $x(t)$ can readily be obtained from the signal's discrete FT $\hat{s}(\xi_n)$ as $x(t) = \sum_{m=1}^{N} \hat{s}(\xi_m)e^{i\xi_m t}$, but this is not so: the discrete FT represents a periodic spectrum estimate, being an exact FT only for signals that repeat themselves with period T. As a result, predicting the signal based on its discrete FT is equivalent to the usual periodic continuation.

There are many methods devoted to fitting the signal with the sinusoidal model [16], using which the parameters c_m, φ_m, ω_m in (7.13) can be estimated reliably (i.e. without the periodicity constraint). However, these methods, although very accurate, are usually quite expensive computationally, as well as giving rise to a variety of issues when applied in practice. Therefore, it is reasonable to use the more convenient and computationally cheaper procedure of estimating tone frequencies by simple iterative fitting of the signal with the corresponding sinusoids, as described below.

Using least squares fitting, the residual error of the signal's fit with $q_0+q_1 \cos\omega t_n+q_2 \sin\omega t_n$ is first minimized over ω; then the value of ω for which the local minimum occurs and the associated "best-fit" parameters $q_{1,2}$ are taken as one of the ω_m, a_m, b_m in (7.13); these two steps are repeated $M-1$ times using in place of signal $s(t_n) \equiv s^{(1)}(t_n)$ the residual signal $s^{(m)}(t_n)$, obtained by subtracting from $s(t_n)$ the tones found at $m-1$ previous steps. To provide more accurate and faster optimization, at each iteration it is recommended to start the search for the optimal ω_m from the frequency of the highest peak in the residual signal's discrete FT $\hat{s}^{(m)}(\xi_n)$.

An important problem of the outlined approach is that the model (7.13) is stationary, and so it cannot adequately describe nonstationary signals, where spectral content changes in time (e.g. when the tones persist only during some time intervals, as in Fig. 2.5a–d). Therefore, for adequate prediction the model (7.13) should reflect mainly a "local" spectrum near the signal's end (t around T). This can be achieved by using a weighted least squares procedure, with more weight being concentrated near the corresponding time boundary. Then the full procedure of estimating the "local" parameters of (7.13) can be summarized as

$$
\begin{bmatrix} q_0^{(m)}(\omega) \\ q_1^{(m)}(\omega) \\ q_2^{(m)}(\omega) \end{bmatrix} = \underset{[\tilde{q}_0,\tilde{q}_1,\tilde{q}_2]}{\operatorname{argmin}} \left[\frac{1}{N}\sum_{n=1}^{N} \mathrm{w}(t_n)\big(s^{(m)}(t_n) - \tilde{q}_0 - \tilde{q}_1 \cos\omega t_n - \tilde{q}_2 \sin\omega t_n\big)^2\right],
$$

$$
\rho^{(m)}(\omega) \equiv \frac{1}{N}\sum_{n=1}^{N} \mathrm{w}(t_n)\left[s^{(m)}(t_n) - q_0^{(m)}(\omega) - q_1^{(m)}(\omega)\cos\omega t_n - q_2^{(m)}(\omega)\sin\omega t_n\right]^2,
$$

$$
\omega_m = \underset{\omega}{\operatorname{argmin}}\left[\rho^{(m)}(\omega)\right],
$$

$$
s^{(m+1)}(t_n) \equiv s^{(m)}(t_n) - q_0^{(m)}(\omega_m) - q_1^{(m)}(\omega_m)\cos\omega_m t_n - q_2^{(m)}(\omega_m)\sin\omega_m t_n, \tag{7.14}
$$

where one begins with $s^{(1)}(t_n) \equiv s(t_n)$, while $\mathrm{w}(t_n)$ denotes the chosen weighting function, and the "best-fit" values of $q_{0,1,2}^{(m)}(\omega)$ are found by weighted least squares for each ω. The search for a minimum $\rho^{(m)}(\omega)$ is started from the frequency of the maximum in the discrete FT of $\sqrt{w(t_n)}s^{(m)}(t_n)$ and, by default, the ω_m for which such a minimum occurs is determined with accuracy $0.01 \times 2\pi/T$. The other parameters of (7.13) are then $a_m = q_1^{(m)}(\omega_m)$, $b_m = q_2^{(m)}(\omega_m)$ and $a_0 = \sum_{m=1}^{M} q_0^{(m)}(\omega_m)$.

Since the maximum amount of nonstationarity (i.e. the "quickness" of the spectrum's changes in time) which can be represented in TFR reliably is determined by the window/wavelet parameters [9], the choice of the weighting function should be based on the window $g(t)$ or wavelet $\psi(t)$ used. Therefore, $\mathrm{w(t_n)}$ is selected as an exponential whose rate of decay is determined from the window/wavelet 0.5-support (2.26), (2.27) in time:

$$
\begin{aligned}
\textbf{WFT: } \mathrm{w}(t_n) &= \exp\left[-\frac{(T-t_n)\log 2}{\tau_2(0.5)-\tau_1(0.5)}\right], \\
\textbf{WT: } \mathrm{w}(t_n) &= \exp\left[-\frac{\omega_{\min}}{\omega_\psi}\frac{(T-t_n)\log 2}{\tau_2(0.5)-\tau_1(0.5)}\right].
\end{aligned} \tag{7.15}
$$

Note, that the time resolution of the WT varies with frequency, so that the optimal $\mathrm{w}(t_n)$ will generally depend on frequency as well. Because using different weighting functions for each ω would be computationally very expensive, the optimal weights for the WT in (7.15) are taken as those for the minimum frequency $\omega_{\min}$, where the wavelet is most spread in time. Such a choice, however, might not be fully appropriate for higher frequencies, so the predictive padding is better defined for the WFT, whose time resolution is fixed (see Sect. 2.2.3).

It remains to choose an appropriate order M of the sinusoidal model (7.13). This can be done using the Bayesian (Schwarz) information criterion (BIC) [15], which was empirically found to be superior to the (corrected) Akaike's information criterion [3, 6, 8] in the present context. Thus, the number M of sinusoidal components in (7.13) can be selected by minimizing the functional

$$BIC = N\log\left[2\pi\rho^{(M)}(\omega_M)\right] + N + (3M+1)\log N, \tag{7.16}$$

where $\rho^{(M)}(\omega_M)$ is given in (7.14), while $3M + 1$ is the number of parameters in the model (one a_0 plus M of each a_m, b_m, ω_m in (7.13)). The order M can be estimated "on the fly" while performing iterations (7.14): at each such iteration one calculates the BIC (7.16), and if two of its consecutive values are higher than the current minimum (which in this case usually coincides with the global minimum), then the procedure is stopped.

The above considerations refer to forecasting the signal for $t > T$. To predict it for $t < 0$, one can use the same procedure on the reversed version of the signal: $s(t) \to s(T - t)$. Finally, it should be noted that the approach outlined does not aim to provide the best or most rigorous predictive scheme, and other forecasting models can be employed instead. Nevertheless, the procedure discussed works well for the majority of signals, being at the same time very straightforward and computationally cheap. It is therefore a convenient choice given that the predictive padding represents a useful addition rather than an essential part of time-frequency analysis.

Remark 7.4.1 For sophisticated signals the order M selected by BIC (7.16) can be quite high, so one should restrict its maximum value to ensure that the cost of the estimation (7.13), $O(MN)$, is not much higher than the computational cost of the TFR calculation itself, which is $O(N_f\tilde{N}\log\tilde{N})$ (with N_f being the number of frequencies at which TFR is calculated, and $\tilde{N}$ denoting the length of the padded signal). To ensure that this is the case, it is reasonable to limit M to the maximum possible number of peaks in the TFR amplitude at each time, which is $N_f/2$. Additionally, since the N data points composing the signal are modeled using $3M+1$ parameters in (7.13), the value of M cannot exceed $(N-1)/3$. Therefore, the maximum order can be chosen as $M_{\max} = \min[N_f/2, (N-1)/3]$.

7.5 Step-by-step WFT and WT Algorithms

MatLab codes for computing the different TFRs considered in this work are available at [1] together with detailed documentation and video-instructions. They allow for the use of any window/wavelet, both standard and user-defined (either in time or frequency domain, or both), and include implementation of all the aspects discussed in previous Chap. 2, e.g. signal preprocessing, automatic determination of the frequency bin widths and the cone-of-influence, different padding schemes (with predictive padding being default) etc. Each algorithm is considered in detail below, summarizing all related issues and steps.

In what follows, it is assumed that the original signal $s(t_n)$ is sampled at $t_n = (n-1)\Delta t, n = 1, \ldots, N$ (so its sampling frequency and the overall time-duration are $f_s = 1/\Delta t$ and $T = (N-1)\Delta t$, respectively), and the signal's TFR needs to be calculated for frequencies $\omega \in [\omega_{\min}, \omega_{\max}]$. In what follows, ϵ will denote the fixed relative precision for determination of the cone-of-influence and padding (see Sect.

2.4.3), while $[x]^{\uparrow}$ and $[x]^{\downarrow}$ will stand for rounding of x up and down, respectively (e.g. $[1.3]^{\uparrow} = 2$, $[1.3]^{\downarrow} = 1$).

The discrete FT of the signal calculated using the FFT algorithm will be denoted as

$$\hat{s}(\xi_n) : [\hat{s}(\xi_1), \ldots, \hat{s}(\xi_N)] = \mathrm{FFT}[s(t_1), \ldots, s(t_N)], \tag{7.17}$$

where usually (e.g. in the MatLab FFT implementation) the FT is returned at frequencies ξ_n in the order

$$\begin{aligned}\big[\xi_1, \ldots, \xi_N\big] = \big[0, 2\pi f_s/N, \ldots, 2\pi[(N-1)/2]^{\uparrow} f_s/N, \\ -2\pi[(N-1)/2]^{\downarrow} f_s/N, \ldots, -2\pi f_s/N\big],\end{aligned} \tag{7.18}$$

so that the positive frequencies come first, and only then the negative ones. The original signal can be recovered by applying the inverse fast Fourier transform (IFFT) algorithm to the signal's discrete FT: $[s(t_1), \ldots, s(t_N)] = \mathrm{IFFT}[\hat{s}(\xi_1), \ldots, \hat{s}(\xi_N)]$. It should also be noted, that throughout this work the literals ω and ξ (e.g. $\omega_{\min,\max}$) denote the circular frequency (in rad/s), with only the sampling frequency f_s, where present, being taken in Hz; in the codes [1], however, all frequencies are in Hz, so that e.g. the frequencies for which the TFR is calculated are returned as $f_k = \omega_k/2\pi$.

Remark 7.5.1 In some steps of the TFR algorithms one will need to calculate the ϵ-supports (2.26), (2.27), as well as ω_ψ (2.14) for the WT. The corresponding formulas assume that window/wavelet functions are known both in time and frequency, but this is not always so, e.g. the explicit form might be available for $g(t)$ only, and not for $\hat{g}(\xi)$. In such cases, one will need to calculate numerically the window/wavelet FT or inverse FT, and use it to obtain the values needed. The implementation of this approach for the general case is quite cumbersome and brings some numerical issues. It will not be further discussed here, but it is included in the codes [1].

7.5.1 WFT $G_s(\omega_k, t_n)$

1. Preprocess a signal as described in Sect. 2.4.1: first subtract a third-order polynomial fit from it, and then bandpass-filter what is left in the frequency band $[\omega_{\min}, \omega_{\max}]$.
2. Select a padding scheme (see Sect. 2.4.3) and accordingly pad the preprocessed signal at both ends with n_1 values to the left and n_2 to the right, where $n_{1,2}$ are determined by (2.36). Denote the padded signal as $s_p(\tilde{t}_j)$ with $\tilde{t}_j = (j-1-n_1)\Delta t$ and $j = 1, \ldots, N_p = N + n_1 + n_2$.
3. Break the frequency interval $[\omega_{\min}, \omega_{\max}]$ into bins $\omega_k = (k - k_0)\Delta\omega$, where $k_0 = 1 - [\frac{\omega_{\min}}{\Delta\omega}]^{\uparrow}$ and $k = 1, \ldots, k_0 + [\frac{\omega_{\max}}{\Delta\omega}]^{\downarrow}$; the optimal frequency step $\Delta\omega$ can be determined by (2.34) (in the latter, $N_b = 10$ is used by default). Note that the positions of ω_k do not depend on any signal parameters (e.g. f_s or T), as is sometimes the case, being determined only by $\Delta\omega$: this is convenient since it

allows WFTs of different signals to be calculated at the same frequencies, so that their characteristics (e.g. mean amplitudes at each frequency) can be compared.

4. Calculate the FT of a padded signal: $[\hat{s}_p(\tilde{\xi}_1), \ldots, \hat{s}_p(\tilde{\xi}_{N_p})] = \mathrm{FFT}[s_p(\tilde{t}_1), \ldots, s_p(\tilde{t}_{N_p})]$, where frequencies $\tilde{\xi}_j$ are given by (7.18) with $N \to N_p, n \to j$. Set the FT at negative frequencies to zero: $\hat{s}(\tilde{\xi}_j \leq 0) = 0$.
5. For each frequency ω_k:
(a) If the explicit form of the window function FT is available (e.g. for the Gaussian window (2.12)), calculate $\hat{g}(\omega_k - \tilde{\xi}_j)$ by direct substitution of the arguments $\omega_k - \tilde{\xi}_j$ into the known $\hat{g}(\xi)$. Otherwise, if only $g(t)$ is available, compute $[\hat{g}(\omega_k - \tilde{\xi}_1), \ldots, \hat{g}(\omega_k - \tilde{\xi}_{N_p})] = \Delta t\, \mathrm{FFT}[g(\tau_1)e^{-i\omega_k\tau_1}, \ldots, g(\tau_{N_p})e^{-i\omega_k\tau_{N_p}}]$, where $\tau_j = -(j-1)\Delta t$ for $j = 1, \ldots, [(N_p-1)/2]^{\uparrow}$ and $\tau_j = (N_p - j + 1)\Delta t$ for other j.
(b) According to the frequency domain form of (2.8), calculate the convolutions $c(\tilde{t}_j)$ of the padded signal $s_p^+(u)$ with $g(u - \tilde{t}_j)e^{-i\omega_k(u-\tilde{t}_j)}$ as: $[c(\tilde{t}_1), \ldots, c(\tilde{t}_{N_p})] = \mathrm{IFFT}[\hat{s}_p(\tilde{\xi}_1)\hat{g}(\omega_k - \tilde{\xi}_1), \ldots, \hat{s}_p(\tilde{\xi}_{N_p})\hat{g}(\omega_k - \tilde{\xi}_{N_p})]$, where $\hat{g}(\omega_k - \tilde{\xi}_j)$ were obtained in the previous substep (a), while $\hat{s}_p(\tilde{\xi}_j)$ were calculated in step 4 (and are zero at negative frequencies).
(c) The WFT at frequency ω_k is then equal to $c(\tilde{t}_j)$ for times within the original signal's time limits: $G_s(\omega_k, t_{n=1,\ldots,N}) = c(\tilde{t}_{j=1+n_1,\ldots,N_p-n_2})$.

7.5.2 WT $W_s(\omega_k, t_n)$

1. Preprocess the signal as described in Sect. 2.4.1: first subtract a third-order polynomial fit from it, and then bandpass-filter what is left in the frequency band $[\omega_{\min}, \omega_{\max}]$.
2. Select a padding scheme (see Sect. 2.4.3) and accordingly pad the preprocessed signal at both ends with n_1 values to the left and n_2 to the right, where $n_{1,2}$ are determined by (2.36). Denote the padded signal as $s_p(\tilde{t}_j)$ with $\tilde{t}_j = (j-1-n_1)\Delta t$ and $j = 1, \ldots, N_p = N + n_1 + n_2$.
3. Break the frequency interval $[\omega_{\min}, \omega_{\max}]$ into bins $\omega_k/2\pi = 2^{(k-k_0)/n_v}$, where $k_0 = 1 - [n_v \log_2 \frac{\omega_{\min}}{2\pi}]^{\uparrow}$ and $k = 1, \ldots, k_0 + [n_v \log_2 \frac{\omega_{\max}}{2\pi}]^{\downarrow}$; the optimal number-of-voices n_v can be determined by (2.34) (in the latter, $N_b = 10$ is used by default). Note that the positions of ω_k do not depend on any signal parameters (e.g. f_s or T), as is sometimes the case, being determined only by n_v: this is convenient since it allows WTs of different signals to be calculated at the same frequencies, so that their characteristics (e.g. mean amplitudes at each frequency) can be compared.
4. Calculate the FT of a padded signal: $[\hat{s}_p(\tilde{\xi}_1), \ldots, \hat{s}_p(\tilde{\xi}_{N_p})] = \mathrm{FFT}[s_p(\tilde{t}_1), \ldots, s_p(\tilde{t}_{N_p})]$, where frequencies $\tilde{\xi}_j$ are given by (7.18) with $N \to N_p, n \to j$. Set the FT at negative frequencies to zero: $\hat{s}(\tilde{\xi}_j \leq 0) = 0$.

5. For each frequency ω_k:
(a) If the explicit form of the wavelet function FT is available (e.g. for the lognormal wavelet (2.19)), calculate $\hat{\psi}^*\left(\frac{\omega_\psi \tilde{\xi}_j}{\omega_k}\right)$ by direct substitution of the arguments $\omega_\psi \tilde{\xi}_j/\omega_k$ into the known $\hat{\psi}^*(\xi)$. Otherwise, if only $\psi(t)$ is available, compute $\left[\hat{\psi}^*\left(\frac{\omega_\psi \tilde{\xi}_1}{\omega_k}\right), \ldots, \hat{\psi}^*\left(\frac{\omega_\psi \tilde{\xi}_{N_p}}{\omega_k}\right)\right] = \Delta t\, \mathrm{FFT}\left[\psi^*\left(\frac{\omega_k \tau_1}{\omega_\psi}\right), \ldots, \psi^*\left(\frac{\omega_k \tau_{N_p}}{\omega_\psi}\right)\right]$, where $\tau_j = -(j-1)\Delta t$ for $j = 1, \ldots, [(N_p-1)/2]^\uparrow$ and $\tau_j = (N_p - j + 1)\Delta t$ for other j.
(b) According to the frequency domain form of (2.13), calculate the convolutions $c(\tilde{t}_j)$ of the padded signal $s_p^+(u)$ with $(\omega/\omega_\psi)\psi^*(\omega_k(u - \tilde{t}_j)/\omega_\psi)$ as: $[c(\tilde{t}_1), \ldots, c(\tilde{t}_{N_p})] = \mathrm{IFFT}\left[\hat{s}_p(\tilde{\xi}_1)\hat{\psi}^*(\omega_\psi \tilde{\xi}_1/\omega_k), \ldots, \hat{s}_p(\tilde{\xi}_{N_p})\hat{\psi}^*(\omega_\psi \tilde{\xi}_{N_p}/\omega_k)\right]$, where $\hat{\psi}^*(\omega_\psi \tilde{\xi}_j/\omega_k)$ were obtained in the previous substep (a), while $\hat{s}_p(\tilde{\xi}_j)$ were calculated in step 4 (and are zero at negative frequencies).
(c) The WT at frequency ω_k is then equal to $c(\tilde{t}_j)$ for times within the original signal's time limits: $W_s(\omega_k, t_{n=1,\ldots,N}) = c(\tilde{t}_{j=1+n_1,\ldots,N_p-n_2})$.

7.6 Fast $O(N)$ Algorithm for Path Optimization

Finding the solution $\omega_p(t)$ to the path optimization problem (3.2) is generally very expensive computationally, usually being carried out by simulated annealing [7]. However, if the functional $F[...]$ depends on the finite number of the subsequent points (rather than the whole ridge profile), then the optimal path can be found in $O(N)$ operations using a dynamic programming algorithm [4, 5], as discussed below.

Consider the functional $F[Q_m(t_n), \nu_m(t_n), \omega_p(t_{n-1})]$, which depends only on the ridge point at the current time t_n (characterized by $Q_m(t_n)$ and $\nu_m(t_n)$) and the frequency of the previous one $\omega_p(t_{n-1})$. Then the optimization problem (3.2) consists of finding the sequence of ridge point indices $m_c(t_n)$ maximizing the integral of this functional over time:

$$\{m_c(t)\} = \underset{\{m_1, m_2, \ldots, m_N\}}{\operatorname{argmax}} \sum_{n=1}^{N} F\left[Q_{m_n}(t_n), \nu_{m_n}(t_n), \nu_{m_{n-1}}(t_{n-1})\right], \tag{7.19}$$

after which the ridge curve is recovered as $\omega_p(t_n) = \nu_{m_c(t_n)}(t_n)$.

It is clear that at each time t_n for each ridge $\nu_m(t_n)$ there exists a unique history of previous peaks $\{\widetilde{m}_c(m, t_n, t_1), \ldots, \widetilde{m}_c(m, t_n, t_{n-1})\}$ which maximizes the integral to this point

$$U(m, t_n) = F[Q_m(t_n), \nu_m(t_n), \nu_{\widetilde{m}_c(m,t_n,t_{n-1})}(t_{n-1})] + \sum_{i=1}^{n-1} F[Q_{\widetilde{m}_c(m,t_n,t_i)}, \nu_{\widetilde{m}_c(m,t_n,t_i)}(t_i), \nu_{\widetilde{m}_c(m,t_n,t_{i-1})}(t_{i-1})]. \quad (7.20)$$

What makes a fast path optimization possible is that, for functionals depending only on the current and previous points, if the profile $\{m_c(t)\}$ maximizing (3.2) includes $\nu_m(t_n)$, then it should include the best path to $\nu_m(t_n)$ as well: $\{m_c(t_1), \ldots, m_c(t_n)\} = \{\widetilde{m}_c(m, t_n, t_1), \ldots, \widetilde{m}_c(m, t_n, t_{n-1}), m\}$. This is because the behavior of $m_c(t_{i=n+1,\ldots,N})$ does not influence the integral over the previously extracted points $m_c(t_{i=1,\ldots,n-1})$. Therefore, at each step one can leave only the best paths to each peak $\nu_m(t)$ and discard all the others.

It is useful to express $\widetilde{m}_c(m, t_n, t_i)$ through the matrix $q(m, t_n)$ which maps the peak number m at time t_n to the previous peak number in such a way that (7.20) is maximized, so that

$$\begin{aligned} q[i](m, t_n) &\equiv \widetilde{m}_c(m, t_n, t_{n-i}) = q(q[i-1](m, t_n), t_{n-i+1}) : \\ q[0](m, t_n) &= m, \\ q[1](m, t_n) &= q(m, t_n) = \widetilde{m}_c(m, t_{n-1}), \\ q[2](m, t_n) &= q(q(m, t_n), t_{n-1}) = \widetilde{m}_c(m, t_{n-2}), \\ &\ldots \end{aligned} \quad (7.21)$$

What remains is to find at each time t_n (starting from t_1), and for each ridge $m = 1, \ldots, N_p(t_n)$, the maximum value $U(m, t_n)$ of the integral up to this point and the index of the previous ridge $q(m, t_n)$ for which this maximum is achieved:

$$\begin{aligned} &\texttt{for}\; n = 1, \ldots, N \;\texttt{do:} \\ &\texttt{for}\; m = 1, \ldots, N_p(t_n) \;\texttt{do:} \\ &q(m, t_n) = \underset{k}{\operatorname{argmax}} \big\{ F[Q_m(t_n), \nu_m(t_n), \nu_k(t_{n-1})] + U(k, t_{n-1}) \big\} \\ &U(m, t_n) = F[Q_m(t_n), \nu_m(t_n), \nu_{q(m,t_n)}(t_{n-1})] + U(q(m, t_n), t_{n-1}), \end{aligned} \quad (7.22)$$

Then $U(m, t_N)$ represents the full integrals (7.19) through the best possible trajectories to each of the last ridges $\nu_m(t_N)$, so that one has $m_c(t_N) = \operatorname{argmax}_m U(m, t_N)$, and the sequence corresponding to this index is the optimal path: $\{m_c(t)\} = \{q[N-1](m_c(t_N), t_N), \ldots, q[1](m_c(t_N), t_N), m_c(t_N)\}$.

For example, for the functional $F[\ldots] = \log Q_{m_n}(t_n) + w(\nu_{m_n}(t_n), \nu_{m_{n-1}}(t_{n-1}))$, which is effectively used in Sect. 3.1.2, the procedure can be summarized as

$$
\begin{aligned}
&t_1 : \texttt{for}\ \ m = 1, \ldots, N_p(t_1) \\
&q(m, t_1) = 0, \ \ U(m, t_1) = \log Q_m(t_1), \\
\hline
&t_2 : \texttt{for}\ \ m = 1, \ldots, N_p(t_2) \\
&q(m, t_2) = \underset{k}{\operatorname{argmax}}\big\{ \log Q_m(t_2) + w(\nu_m(t_2), \nu_k(t_1)) + U(k, t_1)\big\}, \\
&U(m, t_2) = \log Q_m(t_2) + w(\nu_m(t_2), \nu_{q(m,t_1)}(t_1)) + U(q(m, t_2), t_1), \\
\hline
&t_3 : \texttt{for}\ \ m = 1, \ldots, N_p(t_3) \\
&q(m, t_3) = \underset{k}{\operatorname{argmax}}\big\{ \log Q_m(t_3) + w(\nu_m(t_3), \nu_k(t_2)) + U(k, t_2)\big\}, \\
&U(m, t_3) = \log Q_m(t_3) + w(\nu_m(t_3), \nu_{q(m,t_2)}(t_2)) + U(q(m, t_3), t_2), \\
\hline
&\ldots
\end{aligned}
\tag{7.23}
$$

where $q(m, t_1)$ is set to zero because there are no peaks before the starting time.

Numerically, $q(m, t_n)$ and $U(m, t_n)$ represent $M_p \times N$ matrices, updated at each step, where $M_p = \max_n N_p(t_n)$ is the maximum number of peaks; the excess entries $q(\{N_p(t_n) + 1, \ldots, M_p\}, t_n)$ and $U(\{N_p(t_n) + 1, \ldots, M_p\}, t_n)$ are set to Not-a-Numbers (NaNs). Since at each time t_n one needs to calculate for each of the $N_p(t_n)$ peaks the functional with each of the $N_p(t_{n-1})$ of the previous peaks (to find the one maximizing it), the overall computational cost of the procedure is $O(M_p^2 N)$. The outcome of the algorithm is illustrated below on a schematic example:

$$
\begin{array}{rccccc}
t_n = & t_1 & t_2 & t_3 & t_4 & t_5 \\
N_p(t_n) = & 3 & 4 & 5 & 4 & 4
\end{array}
$$

$$
U(m, t_n) = \begin{bmatrix}
1.1 & 1.7 & 2.6 & 3.2 & 3.6 \\
0.8 & 2.0 & 2.3 & 2.9 & \boxed{5.3}\ \text{(max)} \\
1.6 & 2.4 & 3.2 & 4.7 & 5.1 \\
\text{NaN} & 1.9 & 3.6 & 3.8 & \text{NaN} \\
\text{NaN} & \text{NaN} & 2.8 & \text{NaN} & \text{NaN}
\end{bmatrix}
$$

$$
q(m, t_n) = \begin{bmatrix}
0 & 1 & 1 & 1 & 2 \\
0 \leftarrow & 2 & 1 & 2 & \boxed{3} \\
0 & 2 & \nwarrow 2 \leftarrow & 3 \swarrow & 3 \\
\text{NaN} & 3 & 3 & 4 & \text{NaN} \\
\text{NaN} & \text{NaN} & 3 & \text{NaN} & \text{NaN}
\end{bmatrix}
$$

$$
\begin{array}{rccccc}
m_c(t_n) = & 2 & 2 & 3 & 3 & 2
\end{array}
$$

Note, that in this example there are two ways of going from the second peak at time t_1: either to the second row ($m_c(t_2) = 2$), corresponding to $U(2, t_2) = 2.0$, or to the third one, corresponding to $U(3, t_2) = 2.4$. Naively, one would select the third peak, but the path optimization approach (3.2) explores all the possibilities, and finds out that going through the second peak leads at the end to the higher path functional.

The algorithm outlined for optimizing (7.19) can be adapted for functionals depending on any finite number of consecutive ridge points. However, the longer the history that one needs to take into account, the more computationally expensive it becomes. For example, if the functional $F[...]$ depends on two previous points $\omega_p(t_{n-1})$ and $\omega_p(t_{n-2})$, then one can apply the same procedure but instead of single ridges treat their one-step sequences. Thus, in this case one selects the trajectory maximizing the path functional (3.2) to each of the $N_p(t_{n-1}) \times N_p(t_n)$ point combinations $\{\nu_k(t_{n-1}), \nu_m(t_n)\}$. The general case of accounting for d previous points is qualitatively similar, so the computational cost of the procedure is $O(M_p^{d+1}N)$, with the required memory of $O(M_p^d N)$.

References

1. The NMD Toolbox for MatLab, which contains all the needed codes, is freely available at http://www.physics.lancs.ac.uk/research/nbmphysics/diats/nmd/ (see also http://www.physics.lancs.ac.uk/research/nbmphysics/diats/tfr/)
2. P.S. Addison, *The Illustrated Wavelet Transform Handbook: Introductory Theory and Applications in Science, Engineering, Medicine and Finance* (IOP Publishing, Bristol, 2002)
3. H. Akaike, A new look at the statistical model identification. IEEE Trans. Auto. Cont. **19**(6), 716–723 (1974)
4. R. Bellman, *Dynamic Programming* (Princeton University Press, Princeton, 1957)
5. D.P. Bertsekas, *Dynamic Programming and Optimal Control* (Athena Scientific, Belmont, 1995)
6. K.P. Burnham, D.R. Anderson, *Model Selection and Multi-model Inference: A Practical Information-Theoretic Approach* (Springer, Berlin, 2002)
7. R.A. Carmona, W.L. Hwang, B. Torresani, Characterization of signals by the ridges of their wavelet transforms. IEEE Trans. Signal Proc. **45**(10), 2586–2590 (1997)
8. C.M. Hurvich, C.-L. Tsai, Regression and time series model selection in small samples. Biometrika **76**(2), 297–307 (1989)
9. D. Iatsenko, P.V.E. McClintock, A. Stefanovska, Linear and synchrosqueezed time-frequency representations revisited: overview, standards of use, reconstruction, resolution, concentration, and algorithms. Dig. Signal Proc. (in press) (2015). doi:10.1016/j.dsp.2015.03.004
10. G. Kaiser, *A Friendly Guide to Wavelets* (Birkhäuser, Boston, 1994)
11. J.M. Lilly, S.C. Olhede, Higher-order properties of analytic wavelets. IEEE Trans. Signal Proc. **57**(1), 146–160 (2009)
12. J.M. Lilly, S.C. Olhede, Generalized Morse wavelets as a superfamily of analytic wavelets. IEEE Trans. Sig. Proc. **60**(11), 6036–6041 (2012)
13. S. Mallat, *A Wavelet Tour of Signal Processing*, 3rd edn. (Academic Press, Burlington, 2008)
14. S.C. Olhede, A.T. Walden, Generalized Morse wavelets. IEEE Trans. Signal Proc. **50**(11), 2661–2670 (2002)
15. G. Schwarz, Estimating the dimension of a model. Ann. Stat. **6**(2), 461–464 (1978)
16. P. Stoica, R.L. Moses, *Spectral Analysis of Signals* (Pearson/Prentice Hall, Upper Saddle River, 2005)

MIX
Papier aus verantwortungsvollen Quellen
Paper from responsible sources
FSC® C105338

If you have any concerns about our products,
you can contact us on
ProductSafety@springernature.com

In case Publisher is established outside the EU,
the EU authorized representative is:
Springer Nature Customer Service Center GmbH
Europaplatz 3, 69115 Heidelberg, Germany

Printed by Libri Plureos GmbH
in Hamburg, Germany